POPULATION AND THE ENVIRONMENT

Encyclopaedia of Environmental Pollution and Awareness in 21st Century Series

POPULATION AND THE ENVIRONMENT

Edited by

V.K. Prabhakar

ANMOL PUBLICATIONS PVT. LTD.

NEW DELHI - 110 002 (INDIA)

ANMOL PUBLICATIONS PVT. LTD.
4374/4B, Ansari Road, Daryaganj
New Delhi - 110 002

Population and the Environment

First Edition, 2001

ISBN 81-261-0924-6

PRINTED IN INDIA

Published by J.L. Kumar for Anmol Publications Pvt. Ltd., New Delhi - 110 002 and Printed at Mehra Offset Press, Delhi.

CONTENTS

PREFACE

Today, the cry of 'pollution' is heard from all the nooks and corners of the world, and pollution has become a major threat to the very existence of mankind on this earth. It is the major challenge of our times.

The pollution of various resources has gone to such an extent that we are unable to breathe fresh air and drink fresh water. On one hand the advancements of science and technology have added to the human comforts by giving us automobiles, electrical appliances, supersonic jets, space crafts, better medicines, better chemicals to control harmful insects and other pests, etc., but, on the other hand, they have given us a very serious problem to face that is pollution. Pollution can be defined as an undesirable change in the physical, chemical, or biological characteristics of the air, water, or land that can harmfully affect health, survival, or activities of humans or other living organisms.

Pollution does not have to cause physical harm, but it may merely interfere with human activities. Chemicals spewed into the air or water from an industrial plant may be harmful to humans and other organisms living near the plant. If expensive pollution controls are required, the plant may be forced to shut down.

India faces serious problems of environmental pollution. Water pollution from untreated community and industrial wastes flowing into our rivers and streams, increasing use of fertilizers and pesticides, dumping of organic wastes etc., are issues that need to be tackled on the priority. Air pollution, long regarded as the bane of industrialised nations looms large over our country.

Here, we are presenting this *Encyclopaedia of Environmental Pollution and Awareness in Twenty-first Century* in fifty volumes with a view to make people conscious about the seriousness of pollution in the present time and its remedies.

Editor

1 Introduction

A population is a group of living individuals set in a frame that is limited and defined in respect to both time and space. Also population is an interacting, or potentially interacting group of individuals usually of the same species. In the initial study of populaion it is helpful to recognize certain attribute of population which distinguish them from individual organisms like organisms have a structural organization, a functional unity, and a pattern of growth and development unlike individuals they have group attributes and statistical properties that no single individual possess. Genetic characteristics especially applicable to population ecology as for example, adaptiveness reproductive fitness and persistence. Individual organisms are born, have age, grow and die but such characteristic as birth rate, death rate, density and other listed above are meaningful only at the group level. If we are to understand thoroughly the ecology of a species, we must study and measure these population group characteristics as well as know the life history and identifying features of the species. And of course, we all must be concerned about the vital statistics of human population.

Some of the most important aspects of population structure are population size, density, spatial distribution age and sex rates, breeding structure and social organization. Population sizes may vary from a few individuals to millions of individuals. The population of whooping cranes (Grus American) numbers about 50 individuals a single group of birds barely clining to existence. Other rare and endangered species around the world also represent very small populations. The world's population of the Indian lion (Panthera Ieo) numbers less than 200 individuals living in the Gir forest of western India. The California condor (Gymonogyps californicus), golden marmoset (Leontideus rosalia) Javan rhinoceros (Didermocerus sumatrensis) orangutan (Pongo pygmaeus), Gray jungle fowl (Gallus sonnerati) and many other species are now represented by small remnant populations wavering on the edge between life and death. At the other extreme, the world contains many species of tremendous abundance. A population of starlings in a sigle winter flocking roost in eastern United States may number 5 million birds. The white-tailed deer population of Pennsylvania was estimated in 1946 to contain 1 million deer; the rhesus monkey population of one province in northern India was estimated in 1960 to contain between 800,000 and 1 million monkeys (it now probably contains less than half that number); the Norway rat population of Baltimore was estimated in World War II at 400,000; and the bison population of the Great Plains was estimated to contain 60 million individuals in the early nineteenth century, though by the 1890's the bison population had crashed to less than 600 through the greed and disruptive influence of man.

These figures raise the question; of course as to how such data are obtained, and this is a large and complicated subject in itself. Censusing wild animal population is fraught with many difficulties of sampling error and statistical bias. Occasionally direct counts are possible as in aerial photographs of big game herds or sea bird colonies, but for most animals this is impossible and various kinds of sampling methods must be employed. Most animals are not readily visible because of their behavior and habitat, or because they exist in such abundance or scarcity that they cannot be readily counted.

It therefore becomes necessary to estimate numbers through programs of capture and resembling. With many animals systematic program of capture-mark-recapture has been helpful in obtaining estimates of abundance. The ratio of marked to unmarked animals in subsequent trapping runs provides a population estimate known as the Lincoln Index.

If, for example one is studying a resident bird or rodent population in a forest, and succeeds in capturing and banading 100 individuals all of which are released, and then if on a second trapping program, one again catches 100 animals of which 20 were previously banded one would estimate the total original population at 500 animals.

Population density is an important aspect of both population size and distribution. In many populations, the precise boundaries of the population are unknown. Though the distributional limits of the species may be recognized each species usually consists of several populations. The essential feature which distinguishes a population is whether or not the individuals are potentially interactive, or potentially interbreeding. If a species consists of two or more discrete aggregations, which do not interact or influence each other these aggregations can then be considered populations. If this condition persists over an extended period of time, different races, subspecies and even species may emerge as gene frequencies change and evolution occurs. Since populational boundaries are often unknown the most practical description of population abundance is frequently in terms of densities. As an example many ecologic studies of small rodents have shown population densities usually in the range of 0.5 to 20 animals per acre. But (1940) showed population densities of the deer mouse (Peromyscus leucopus)in southern Michigan woodlots to range from 3.08 to 10.32 individuals per acre, Blair (1951) estimated the population density of beach mice or old field mice (Peromyscus polionotus) to range from 0.83 to 1.41 individuals per acre. Ashby (1967) estimated population densities of the bank vote (Clethrionomy glaeolus) a relative of the meadow mouse, to run as high as 17 to 18 individuals per acre in central England.

An important consideration, which will be emphasized later, is that many populations fluctuate widely in density and size in the course of natural events. Occasionally small mammal populations reach amazingly high densities. In California, Krebs (1966) described a meadow mouse or vole population (Microtus Californicus) which achieved a density of 150 to 300 individuals per acre in a relatively small area, and Parson (1963) described a population of house mice in California which increased from a density of less than 5 mice per acre to 300 mice per acre within 6 months. At these high levels, or course, rodents can become agricultural problem and can precipitate economic losses of crops or stored food products. Splitte (1968) has estimated that wild rats (Bandicota bengalensis) in Calcutta grain storage warehouses may achieve densities of one rat per sequare meter, equivalent to 4,000 per acre. These represent extreme concentrations, or course, but they indicate the population potential of some species.

Another important consideration in concepts of both population size and density is the spatial distribution of individuals. Spatial distribution involves a fascinating interplay of behavioral and ecologic factors, and is a prime example of the inseparable tie between behavior and ecology.

The reason for clumping or nonrandom distribution might be numerous. In some cases clumping would be attributable to the heterogeneity of the environment the distribution of food, cover and shelter. For example, dung beetles in an old field would be clumped around the dung isopods and millipedes around the fallen log, etc. In other cases, the clumping might be due to the social behavior. This clumping may be related to environmental resources, but not necessarily so.

This is not to say that random dispersion of animals never occurs. Park pointed out that flour beetles (Tribalism confusum) are distributed in flour according to a Poison distribution, indicating that the individuals are spaced at random. Cole found that certain spiders were distributed randomly under boards placed on the forest floor. These examples are probably exceptions to the general rule, and they are most likely attributable to either the homogeneous or artificial nature of the environment in each case. For purposes of generalization it can be stated with reasonable assurance that most animals and many plants in natural environments show a clumped or nonrandom pattern of distribution.

It is obvious that animals within a population differ in a number of ways: sex, age, breeding condition health, physical condition and social status. Demography, or the numerical analysis of populations, is concerned with the sex, age and breeding condition of the population. Medical science and epistemology are primarily concerned with the health and physical condition of the population and behavioural science is primarily concerned with the social and behavioral states within the population.

In most vertebrate populations primary sex ratios at hatching or birth approach 50 percent male and 50 percent female. In some forms, there is a slight deviation from this; in man, for example sex ratio at birth is closer to 52 percent male, and this is true for rabbits cattle and many birds. In domestic chickens, sheep and horses, sex ratios at birth or hatching are usually closer to 49 percent male. The secondary sex ratios or adult sex ratios of many vertebrates show greater diviations, however. In the Alaskan fur seal one adult bull may dominate a breeding group of 30 females. Wheras the total adult male sex ration is greater than 3 percent the excess adult males are on the periphery of the breeding grounds, they are essentially surplus, and they suffer substa-ntially higher mortality. In most primate populations, adult females out number adult males two or three to one. Apparently males have higher mortality rates throughout life than females have higher mortality rates than males, particularly during the nesting season. In waterfowl and gallinaceous birds, for example male often outnumber females. The sex ratio of blue-winged seal (Querquedula discors) in fall and winter flocks averaged 59 percent males in a sample of 5,090 birds and in the early summer breeding population of mallards (Anas platyrhynchos) and pintail (Anas acuta) in Manitoba, males may outnumber females 4 or 5 to 1.

Age ratios represent another important element of population structure, and one that is also particularly valuable in analyzing populations dynamics shows age structure of populations in terms of three basis types: (A) a declining population, with a low percentage of young in the population (B) a stable population with a large percentage of young than adults, and (C) an increasing population, with a very large percentage of young. The exact shape and proportions of these age pyramids are a function of natality, mortality and population turnover. Some populations characteri-stically have tremendous production of sex cells and youngs with massive morality of young. The Pacific herring, for example produces 8,000 eggs per famale per season, of which 95 percent hatch but only 0.1 percent survive to maturity. The female shad lays from 30,000 to 100,000 eggs per female per season, and less than 0.1 percent grow into mature fishes. Populations as such would have age pyramids with very broaad bases with tiny segments ascending into adult levels. At the other extreme, elephants and whales usually produce only one young per female every few years, and they would have age pyramids with relatively narrow bases and expanded adult proportions. That is, the majority of animals in the population would be adults.

Rhesus populations in India show different age structures in different habitats. In roadside habitats and villages, where juvenile rhesus monkey (1 to 3 year of age, postweaned but preadult) are trapped for export for use in biomedical research and pharmaceutical production throughout the world, there is a conspicuous juvenile "gap" that is, an unnatural shortage of these individuals. These populations have been declining throughout the 1950's and 1960's at the rate of approximately 5 percent per year, as India has exported any where from 40,000 to over 100,000 rhesus monkey per year, primarily juveniles. In temples where rhesus monkeys are protected to a certain extent for religious reasons, and in forests where they can escape trapping more successfully age structure of an island population of rhesus monkeys on Cayo Santiago Island off the eastern coast of Puerto Rico is also shown. This population has been fed and protected by a governmental research program of the National Institutes of Health, and it shows a broad base of infants and juveniles. This population was increasing at the rate of 16 percent per year in the early 1960's and gave a good indication of the natural tendency of primate populations to increase if protected in a satisfactory habitat.

The question of population has occupied the attention of state men and philosophers since ancient times. But it is only recently that attempts have been made to investigate systematically the cause of population change and the specific ways in which population dynamics influence human welfare. Plato and Aristotle were interested in the question of population size in the context of a city-state. To them the optimum size of the population was that is with which man's potentialities could be fully developed and his "highest good" realized. Thus was possible when the population range was enough to be economically self-sufficient and capable of defending itself, but not too large for constitutional government. Plato specified 5,040 as the number of citizens "most likely to be useful to all cities".

European authors on population during the early modern and the medieval periods favoured a growing population. The discovery of the New World, the increase of commerce between Europe and Asia, and the rise of national state brought about some changes in the terms of discussion of population questions, but not until the later part of the eighteenth century was there any marked shift in the general attitude favouring large and increasing population. The mercantilist and cameralist schools of political economy, which flourished in Europe during much of the seventeenth and the eighteenth century, emphasized the economic, political and military advantages of large and growing population and favoured various measures to stimulate population growth. These writers were concerned primarily with the ways and means of increasing the wealth and power of the State. Their aim was not to promote per capita income but to increase the aggregate national income which was viewed as a sources of revenue for the State.

During the last half of the eighteenth century, a number of writers on economic and social questions rejected the mercantilist doctrine and the view that population growth was advantageous and should be actively encouraged by the State. Some writers, specially in England, France and Italy, emphasised the dependence of population upon means of subsistence. They opposed arrangement for poor relief on the ground that it might undermine frugality, make labour immobile, depress productivity and thus increase the pressure of number on means of subsistence. These arguments were turned against such advocates of social reform as Godwin and Condorect, in an effort to show that any benefits from reform would be cancelled by a consequent increase of populations.

At the close of eighteenth century, Thomas Malthus, an English economist, noted this tendency for populations, including human population to increase rapidly. In 1798, he published his famous work, Essay on the principal of population, in which he put forth the idea that populations tend to increase faster than their means of subsistence. In general, he felt that populations tended to increase geometrically or exponentially, whereas their food supplies and means of subsistence tended to increase only arithmetically. A geometric increase is a series of numbers having a common ratio, such as 2...4...8...16...32...64 in which each succeeding number is twice the former. An arithmetic increase is a series of number having a common difference, such as 2...4...6...8...10...12, in which each number differs from the other by 2. Thus, Malthus thought that populations constantly tended to outstrip their food supplies and were then decimated by poverty, starvation, disease, warfare, or other catastrophes. Malthus was primarily concerned with human populations, and he felt that moral restraint was the solution to human population pressures.

Malthus's theory is thus based on two postulates and one assumption. His postulates are that (i) food is necessary for the existence of man, and (ii) that passion between sexes is necessary, and will remain nearly in its present form. The assumption is that the production of food increases in arithmetical progression while the population increases in geometrical progression. Thus the power of population is indefinitely greater than the power of the earth to produce subsistence for men. Efforts must, therefore, be made to keep the population from growing beyond the means of subsistence through some powerful checks, else it will lead to "vice and misery."

Malthus talked of two kinds of checks to population positive and preventive. He termed preventive check as "prudence" and included in it postponement of marriage and restraint on reproduction. Positive checks he termed as "natural" as "they arise form the situation itself" and included in it wars, pestilence, epidemics, diseases, famines, natural calamities and the like.

Malthus was highly praised and also bitterly criticised for his views. His Essay aroused a storm of controversy which long out-lived Malthus himself, and which made both his followers and his opponents conscious of the need for adequate information on population trends and their influence on social and economic conditions. Thus Malthus was indirectly responsible for developing population censuses a vital statistics. Malthus has been criticised for his postulates, his assumptions and his conclusions. While it is not to be disputed that food is necessary for man, but so is water clothes, shelter, and such other prime necessities of life.

There is enough historical evidence to indicate that the output of food and of other necessities has increased faster than population, and has been brought about without any distinct symptoms of increasing "misery and vice", but rather of their diminution. With improvement in techniques of production increasing returns are being obtained from land and from the production of prime necessities of life.

Malthus has proved a false prophet. While he postulated great miseries on account of unprecedented growth of population, it is found that birth rate in most of the advanced countries of the world has declined, and that the rapid growth of population in under-developed countries is basically due to a rapid fall in death rate and not on account, as Malthus feared, of an increase in birth rate.

Marx rejected the Malthusion theory of population. He held that there is no universal law of population, and that the source of "over-population" was not to be found in man's biological power of reproduction, but in prevailing capitalistic mode of production. Over population arose because capital accumulated less rapidly than did the supply of labouring population was not only a necessary consequence of capita-listic accumulation, but also a necessary condition for the existence and continuation of the capitalistic form of economic organisation.

The realisation that Malthus has over-generalised a particular case lead to a re-thinking of the population question. Writers at the beginning of the present century began thinking that population growth is not always undesirable. This led to the development of the optimum theory of population which is associated with the name of Professor Cannan and others.

The theory says that given the natural resources and the technique of production, that population, Cannan envisaged a point of maximum return for all industries taken together, that is the population which will maximise productivity. He also indicated that the optimum size of the population changes as circumstances change and as new techniques of production are introduced.

Effort to precisely define the "optimum" led to the refinement of the concept, but at the same time to the theory being criticised as of little practical value. It is considered highly artificial to select "average real income "as the soletest of the optimum. In fact there can be an optimum size of the population from ecological, political and other points of view. Again, the optimum size is subject to frequent changes on account of changes in techniques of production.

The transition theory, development in the 30's of third century and associated with names of Warmn S. Thompson and Frank W. Notestein, explains the relationship between population with and economic development. It says that in developing countries, which are characterised by heavy pressure of population, low productivity heavy dependence on agriculture, rudimentary techniques of production, under-developing means of transport, and poor sanitary conditions, birth and death rate high. Death rates are high because of poor diet, primitive sanitation, and the absence of preventive and curative medical facilities. Birth rates are high because of social beliefs and customs which encourage large families and which group up if a high death-rate community is to continue its existence. The birth rate around 40 per one thousand person and the death rate around 35 so that the population does not grow at a fast rate.

With economic progress the death rate begins to decline. This is because of improved means of transport, better sewage disposal and improvement in drinking water facilities. But the birth rate stays high so that the difference population tend to grow at the rate of 20-30 per thousand population per year. Because of the rapid rate at which the population grows, this period is also known as that of "population explosion".

One of the features of economic development is typically increasing urbanisation and in an urban setting children are usually more of the burden and less of an asset. The processed economic change also weakens the force of traditional customs and beliefs. Females are being to realize that if burdened with a large family they are unlikely to play their due role in society. Consequently, the idea of large family yields place to the small-family-death and the birth rate declines from the high 40 to about 17 per one thousand population. The death rate is also low, about 8 per one thousand, so that the population just maintains itself.

The three stages described above are also called the stages of slow population growth, rapid population growth and of stable or declining population growth. The industrially advanced countries of the world have passed through the first-two stages and are current in third stage. Developing countries of Asia, Africa and Latin America are either in the first stage or entering the second stage.

The sequence of events described by the theory can be traced in every region when the economy has evolved from an agrarian to an industrial market oriented economy. But it is not specific to quantity if the degree of decline of the vital rates. The theory, however, contains one significant generalisation, namely, the decline in birth rate occur after a substantial time-lag in comparing with the decline in death rate, and that during this lag population increases very rapidly. For instance, the "area of European settlement" increased its population six fold during the period 1750 and 1950. The population more than doubled in the century from 1750 to 1850 and nearly tripled in the period 1850 to 1950.

The theory is of considerable significance to under developed countires of Asia, Africa and Latin America. Their birth rate is high and the field of public health, it has been possible for the death rate to be substantially reduced with marked changes in the economy or in the birth rate., Consequently, their population is growing at a rate which doubles itself in twenty to twenty-five years. This is tending to retard economic progress. There is therefore an urgent need for bringing about a substantial decline in birth rates if economic development is to become a continuing process.

The first major scientific challenge to Malthusian doctrine as a pattern of population growth appeared in the 1830's in the work of Pierre Verhulst. In 1839 Verhulst proposed that population normally grow in a much more orderly fashion than that proposed by Malthus, describing in their growth a curve of S-shaped proportion. This became known as the Logistic Theory of population growth. The logistic theory asserted that population have a slow initial growth rate, which increases exponentially unitl it reaches a maximum, and then becomes progressively less as the population approaches an upper limit of its growth. The upper limit is approached gradually and in an orderly, predictable manner. The resulting curve is the familiar S-shaped of population growth.

It should be noted that the logistic curve and the Malthusian or irrupting curve do not differ in the early stages of population growth—both show a slow start followed by a period of exponential or geometric growth but they do differ fundamentally in the upper or controlling stages of growth. The Malthusian curve is characterized by an erratic, often catastrophic, pattern of limiting growth, whereas the logistic curve characterized by a smooth, orderly and gradual pattern.

Following the work of Verhulst, the logistic theory lay dormant unto it was independently derived and popularized by Raymond Pearl in the 1920's. Pearl applied the logistic curve to the population growth of yeast, Drosophila and man. This ushered forth a burst of activity resulting in the application of the logistic curve to the population growth of protozoa, water fleas, pond snails, thrips, ants, bees and other organisms by many investigators.

With this wealth supporting evidence, the logistic thoery attained wide acceptance. It was proclaimed as a law of population growth by some biologists and was used to predict future population levels of experimental and natural population. Pearl's investigation led him to conclude: "It has been shown in what has preceded that populations of living thing 'from the simplest' represented by yeast, to the highest, represented by man, grow in accordance with the logistic curve. One can now feel more certain that this curve is a first and tolerable close approximation to a real law of growth for human populations.... We can, for example, upon a more adequate scientific basis mere than guess-work, predict future population, or estimate past populations, outside the range of known census counts."

That some biologist accepted Pearl's interpretation rather freely is indicated by statements of Clarke: "Population of a wide variety of organisms, ranging from bacteria to whales, have been found to follow the logistic curve in the growth from.... The growth of man's population follows a similar pattern whether examined in individual regions or in the world as a whole. "Clarke presented a graph of population growth in the United States fitted with a logistic curve, and observed an upper symptote of 184 million in year 2100 A.D.

Many other biologists and mathematicians, however, severely criticized the significance of the logistic theory. In fact, the theory met stern opposition since its first expression could be made to fit a wide variety of curves, often of very different nature. Gray, Hongben, and Wilson and Puffer registered doubt as to the universal validity of ligistic theory and commented on its inadequacies as a description. Feller pointed out that the actual fit of the logistic curve to much of the published data was poor and better agreement could be obtained with other formulas. Andrewartha and Birch and sang accurately pointed to limitations of the logistic equation. Sang in careful reexamination of population growth in Drosophila, emphasized that, "The data we have summarized show that only in very exceptional circumstances would logistic growth occur in a Drosophila culture... The ecological situation is too complex to be adequately described as the Pear-Verhulst law, "Sang continued to discuss general aspect of the logistic theory by stating, "This acceptance of a plausible formula, in spite of its inaccuracy, may be justified; provided this attitude does not inhibit future research, particularly in to fundamental conception. In the case reviewed, it seems that wide acceptance of the logistic law has led to just this kind of inhibition of further work, and that many interesting ecological and physiological processes have been ignored as a result.

The study of vertebrate population in both natural and confined circumstances has produced some population growth curves deviating considerably for logistic growth and tending to ward errupting or

Malthusian growth. The sheep population of southern Australia showeed a pattern of liogistic-like growth from 1840 to 1890, but its upper level was reached abruptly and was followed by substantial mortality. The deer of Kaibab National Forest in Arizona grew from an estimated population of 4,000 animals in 1918 after a period of extensive predator control to an estimated population of 1,00,000 in 1924 but in the winter of 1924, they died by the thousands after the population had destroyed its food supply. Their growth and decline described a classic Malthusian curve.

The pheasant population of Protection Island, Washington, showed a phenomenal increase from 8 birds on the 397 acre island in 1937 to nearly 2,000 birds by 1942. In one year the population showed an increase of 3000 percent. Unfortunately the field research on this population was terminated before natural control occurred, but the population seemed to be describing a typical irrupting population curve. Several studies of seminatural populations of wild house mice have also shown irrupting population curves.

Seminatural populations have been maintained in confined, but free-ranging conditions with a natural habit for this species, excep' for the element of confinement. Various factors such as space, food supply, resting cover and the structure of the physical environment have been experimentally controlled to elucidate their roles in population growth and limitation Work of this type at the University of Wiconsin and the Institute of Ecology in Warsaw Poland, has shown that population growth in *house mice* is highly variable, and often dependent upon patterns of sicial behaviour within the population. For example, populations in identical environments showed widely variable patterns of population growth. The variability was related to social behaviour within the population in terms of aggressive behavior, dominance hierarchies, sexual behaviour and parental care. All populations become controlled even though abundant food and nesting space was still available when aggressive behavior increased to an abnormal degree, when the nesting territories of females were no longer properly maintained, and when parental care of the nestlings become aberrant. Infant mortality increased through nest desertion and cannibalism.

The densities at which these behavioral pathologies occurred were highly variable between populations, and they were function of the crowding tolerance and behavioral stability of individuals. This complicated interplay of ecology and social behavior was not adequately accounted for by the logistic theory and the resulting population curves were totally individualistic and unpredictable. A fairly complete evaluation of logistic theory has been presented by Andrewartha and Birch. They have analyzed the inherent assumptions of the logistic equation and have reviewed many studies in the light of these assumptions. The analysis has led them to conclude that the logistic curve provides an inperfect, yet useful, description of the population growth of organisms with simple life histories (such as yeast and Paramecium); but for animals with complex life histories (such as insects and vertebrates) the population growth usually does not conform closely to the logistic curve except for small parts of the data.

The major question in human terms is, of course, what pattern of population growth characterizes man? From the early seventeenth century to the present time, the world's population has displayed exponential growth. There seems to be widespread faith that human populations shall ultimately reveal a logistic pattern, but certainly all attempts to predict human population levels by means of the logistic model have been very far from the mark. Pearl predicted a world population of 2.6 million by the year 2100. This population level was exceeded many years ago, and economic planning based on Pearl's prediction would indeed have been disastrous. Although we continue to hope that human population throughout the world will achieve a pattern of graduallogistic balance through orderly and intentional process, there is no sound scientific evidence that this will in fact occur. The possible of catastrophic mortality through war, disease and starvation are still very much with us. So also are other patterns of mortality which are not necessarily catastrophic in their total effect, but are clearly related to density, crowding and modern population pressures. In this category are automobile accident mortalities, which are responsible for 50,000 deaths per year in the United States alone, various stress-related diseases, including coronary artery disease and high blood pressure, and various pollution-related diseases including emphysema and chronic bronchitis. If deaths attributable to these causes continue to show present trends, these mortality factors will begin to play a more significant role in population regulation, especially in technologically advanced countries. Different patterns of population growth may be exhibited by populations recovering from major depletions, or by populations invading new inhabitants. Since most populations are already established, however, and are not growing de novo, they are instead showing varying degrees of fluctuations.

Our familiar concept of the balance of nature implies stability but the fact is that most animal populations under go continual change. Natural populations often exhibit a dynamic resiliency, rising and falling in response to many factors. In one sense, a pattern of population fluctuation such as that normally exhibited by most species, may be considered continuous series of growth & decline sequeuces. Thus study of population fluctuation becomes closely allied with the study of population growth and natural control.

At the outset it is convenient to separate population fluctuations which are seasonal (i.e., obviously related to seasonal weather), from those which are nonseasonal latitudes, most animal pupulations have breeding seasons so that population growth typically occurs in the spring and summer seasons.

Most arthropods and vertebrates begin producing young in the spring and summer seasons, so that these seasons are characterized by population growth, and they cease the production of young in laten autumn and winter. In aquatic ecosystems, many populations also show marked seasonal fluctuations. Both phytoplankton (miscoroscopic algae), and zooplankton (protozoa rotifiers, small crustacea, etc.) usually show spring and fall increases in population. These sharp increases, often 10 to 20 fold, are known as "pulses". They are sometimes correlated with temperature changes or with turnovers in the water state which recirculate nutrients.

In the tropical regions of the world, where there are not sharply defined seasons on a temperature basis, reproductive seasonality still occurs in many plants and animals, and it is often related to the cycle of wet and dry seasons. Many tropical insects, for example, have sharp peaks of abundance related to the monsoon seasons. Even vertebrate animals, including primates, have been found in nature to have definite periods of breeding and birth seasons, often correlated with the monsoons.

In both tropic and temperate zones an important principle seems to be that the young appear at the most favourable time of the year in term of food and climate. Thus young deer and antelope are born when insect food is most readily available. These are exceptions to this general principle; for example, young rhesus monkeys are born in the midst of the hot dry season in northern India, when temperatures are excessively high (often over 100 degree F), water is extremely scarce, and plant growth is virtually nil. In this case, however, the infant feeds exclusively on the mother's milk for two or three months until the monsoon season starts.

It is possible that natural selection has acted in such a way as to place breeding, conception and the first trimester of pregnancy at the most favorable time of the year, namely, the immediate postmonsoon month when food is most abundant. In other words, the events of conception and early embryonic growth may have been more critical in the life cycle of the rhesus monkey in terms of evironmental influences than the period of early infancy which is well protected by the mother and the social group. This is speculation and not established fact at this time. It also fits with the observation, however, that the monsoon season with its luxuriant growth again arrives when the infant is two or three months old and is relying less on maternal milk and starts taking its own food.

Population fluctuations which are relatively independent of seasons are of two general types: (1) random, or (2) cyclic. Cyclic fluctuations are those which have a definite periodicity, or rhythmicity; that is, the population reaches a peak at fairly regular intervals. In vertebrate populations, certain species reach population peaks every 4 years, while other have a 10-year cycle. These will be discussed later as specialized cases of more generalized random fluctuations.

Ramdom fluctuations may be minor perturbations of fairly stable population, producting relatively flat curves, or they may be major changes in abundance, producing erratic unstable curves. Examples of stable populations are increasingly difficult to find in the modern world. They were probably characteristic of many organisms in undisturbed ecosystems; that is, ecosystems in balance. They were also probably characteristic of complex systems, such as forests and particularly tropical forests where great species diversity produced a complicated web of natural check and balances on each species. Within such a system most vertebrate populations were probably stable and exhibited only their typical seasonal fluctuations.

Probably some of the best modern examples of stable populations may be found among the birds and larger mammals. In birds where considerable species diversity occurs and fairly extensive food resources are available, considerable stability in population size may be present. John Gibb reviewed a number of bird population studies several years ago and noted: "Notwithstanding the considerable annual fluctuation is bird populations that have been recorded in almost every species investigated, it is their comparative stability over long periods that is their most remarkable attribute. "Thus, although wide variations is number occur form year to year in many bird population, these population often show a tendency to return to typical levels.

An example of a relatively stable vertebrate population is afforded by a census study of rhesus monkey populations (Macaca mulatta) in northern India. With complete protection, good health and abundant food supplies, rhesus populations can increase in experimental colonies at the rate of 15 per cent per year. But in India, where monkey experience limited food supplies, varying degrees of protection from the human population, and a considerable number of infections diseases, populations do not show this pattern of increase.

In general, throughout India as in most of the world, primate population are declining due to habitat deterioration (cutting of forest, overgrazing of savannah lands, etc.) human competition (where native people shoot monkeys for food), and high rates of trapping for commercial trade. As an example of the latter, in the late 1950's India exported more than 100,000 monkeys per year for use in biomedical research and pharmaceutical production, though the current figure is only about onehalf of this. These monkeys were essential in the development of polio vaccine, measeles vaccine, and many types of medical and psychological research. The long term population trend of many animals, especially wild vertebrate animals, is down ward.

Since 1600 A.D., 120 species of birds and mammals have become extinct, a number several times greater than the natural rate of extinction. At the present time, approximately one vertebrate animal become extinct on the face of the earth every year, and within the next 30 to 50 years, 100 more species will probably be lost. The International Union of the Conservation of Nature periodically publishes a list of rare and endangered species (the publication is known as the Red Data Book), and it now lists nearly 300 species of birds and 275 species of mammals as endangered.

The peregrine falcon (Falco peregrinus), ospery (Pandion haliaetus), and bald eagle (Haliaectus leuocephalus) are presently among the vanishing animals of North America. Others have come close to extinction, but were saved by intensive conservation movement at the last moment. Throughout the world, the basic causes of extinction are habitat deterioration, poaching, poisoning, and direct or indirect competition with man. From the ecological standpoint, some extinction is an inevitable evolutionary process, but when it becomes excessive as in the last century, it is a serious disruption to biotic communities and ecosystems. Each extinction lessens species diversity, and thereby reduces ecosystem stability. Each estinction also represents the irreplaceable loss of unique biological material. It should be pointed out of course, that some vertebrate animals have increased drastically within the last century, but these are usually domestic animals such as dogs, cats, cattle, goats and so forth, or wild animals that have learned to become commensal with man; that is, animals which have successfully adapted and even exploited the human environment. This includes the domestic rats and mice such as the Norway rat, the house mouse the Asian bandicoot rat (Bandicota bengalensis),the Asian house shrew (Suncus murina) to starling, the house sparrow and the pigeon (Columba livea.)

Even the herring gull (Larus argentatus) is adapting to man's urban environment successfully by feeding on garbage dumps, and gull population on both United States coasts have increased in recent years. Apart from those species which seem to be headed towards a permanent decline and eventual extinction, and those which thrive on man-made environments, the fact remains that most animal population fluctuate substantially from year to year.

It is often a matter of common knowledge among farmers and sportsmen that "this is a good year for pheasants,"or "there were rabbits all over the place last year, but not so many this year," or "this is the world year for mosquitoes we've ever had," and so forth. These are all popular expressions of population fluctuations. Scientifically, these fluctuations are not often recorded in a systematic and continuous fashion, but some data do exist to document the types of fluctuations which are common in animal populations. In trying to manage animal populations, we tend to do the most obvious thing, which would logically seem to produce results, and often does, but in the long run, may be quite detrimental to the total ecological balance of a community.

In controlling insect populations, we use residual poisons such as DDT, but are now discovering after some 30 years, that this has long term side effects which are quite dangerous, and we are also dicovering that many insects have become resistant to it in any case. In controlling the population levels and general health of their prey species. In enhancing the abundance of desirable species, we have often poured money into artificial breeding and restocking programes of animals such as pheasant, wild turkeys and trout only to find that in the meantime, the habitat has so deteriorated through pollution that the transplants have little or no change of success, and the recovery rate per transplant is almost nil. This is not intended as a blanket condemnation of all of our control and management programs, but by and large the balance sheet is not favourable, mainly because we have based our management on trial and error rather than on sound basic knowledge of the ecological factors involved in population growth and regulation.

The most diamatic population fluctuations are those which are cyclic: that is, in which the population reaches peak numbers at fairly regular intervals. Some animal populations, such as those of the snowshoe hare (Lepus americana) have periodicity of abundance of 9 to 10 years.

Other animal populations have a cyclical periodicity of 3 to 4 years. These include the Norway lemming of Scandinavia (Lemmus lemmus), the field mouse (Microtus agrestis) of Europe, and some of the predators of these two notably the colored fox (Vulpes fulva) and the snowy owl (Nyctea scandiaca). Not all populations of these animals always fluctuate cyclically. When they appear to do so, sophisticated statistics may be required for proof of real cyclic periodicity. These animal population cycles have been intriguing and puzzling to biologists for many years. Considerable research has been centered around trying to understand the cause of cycles, and numerous theories have been proposed to account for them.

The final answers are still not clear, though many biological phenomena have been discovered and analyzed in this search for understanding. In general, two types of factors have been considered as the basic causes of animal population cycles:(1) extrinsic factors; that is, those outside of the population itself, such as climate and weather, cosmic events competition between species, predator-prey relationship's and food supply; or (2) intrinsic factors; that is those within the animals as population increases occur and they become crowded.

Several of these factors can be dismissed rather quickly, since extenive research has failed to show any clear relationship. For example climatic and meteorological changes have not been satisfactorily correlated with regular long-term cycles of population change in any animal. Sometimes population peaks coincide with years of great rainfall, or some other notable aspect of weather, but these occurrences are by no means consistent.

At one time, sunspot cycles were considered to be correlated with population cycles. Sunspot maixma have occurred every 10 to 11 years, and they were correlated with peak population of hares and lynx in the late eighteenth and early ninteenthe century, but in the late eighteeth and early ninteenth century, but in the last 100 years, have been entirely out of phase. Changes of populations of predators such as the lynx, fox and sowy owl are clearly correlated with cycles of their prey. But they cannot be shown to cause the prey cycles, for prey cycles often occur in areas devoid of predators. For example snowshoe hare populations have cycles in areas where no lynx are found. There has been some attention given to disease and parastism as the cause of long-term cycles. Twenty years ago there was interest in hypoglycemic shock in snowshoe hares.

Animals at peak periods of abundance seemed to have low blood sugar levels and go into states of shock and death. This was interpreted as the alarm reaction of Selye's general adaptions syndrome. There has also been considerable work on tuberculosis and toxoplasmosis in wild voles and although high disease rates have occasionally been asociated with die-offs of peak population, these associations have not been consistent. Many rodent declines have occourred without evidence of an infectious disease as the cause.

Most ecologists have not been well trained in pathology, bacteriology and virology, not have scientists in these disciplines extended their work satisfactorily to population ecology. Food supply has often been considered inportant in the decline of animal populations. Lack (1954) proposed that food is the main limiting factor in bird populations as it is for deer population in North America, Most field studies of rodents have demonstrated however, that even at peak periods of abundance ample food is present. Chitty (1960) studied grass during and after high vole plagues and found ample food.

There are exceptions; for instance, during lemming peaks food becomes severely depleted and this is one of the factors that is though to start lemming migrations. Pitelka felt that insufficient attention has been paaid to qualitative study of food. The estrogents and steroid in growing plants, particularly grasses and legumes, are now being studied. It has been observed in the Orients, for example, that upsurges of rodent populations are often timed with the flowering and sprouting of bamboo. Then there is not only increased supply, but the young growing plants are more nutritive and estrogens are increased. The estrogens have direct influence on fertility and fecundity of the animals feeding on them.

2 Population Interaction

Even within the same species there are circumstances in which the individuals of some populations are territorial distributed and other circumstance in which they are not. This is true in ungulates for example, many of which are territorial during the rut or mating season, but the same individuals are non-territorial at other times of the year. Apart from this obvious correlation with breeding, however, there is evidence in some species that territorialism is related to the population density and/or environmental circumstances. In the Japanese salmon (Plecoglossus altiveli), individual fish are territorial at low population density, but as density increases, individuals give up their territories and form a school.

Wild house mice (Mus musculus) show similar behaviour with territorialism at low population densities, and mass group behaviour at high population densities. This variability is related to environmental factors and differences in social behaviour of key individuals. In view of these facts, it does not seem warranted to consider territorialism an instinctive pattern of behaviour. Territorialism is certainly common and wide spread, but by no means universal within limited animal groups or even within a single species. It seems more appropriate to consider territorial behaviour a flexible ecological adaptation—a social and behavioural means of utilizing space.

Whereas territories function to allot resources on a spatial basis, dominance hierarchies allot resources on an individual priority basis within the same physical space. Hierarchical systems have been observed in all vertebrate classes and in many invertebrates. Dominant animals have priority access to food, mates, nest sites and resting locations. Dominance is often manifested between animals by single physical displacement by the displacement of one individual by another at a feeding or resting site. Occasionally it involves displays or threats, and rarely it may involve direct fighting. The latter is more common in initial encounters in which the dominance rank or social status of each individual is not yet determined, but once established the hierarchy is then usually established by display or by social memory.

Once a rhesus monkey has established high rank, it does not have to constantly display this rank in a stable social group. The high rank of some individuals and low rank of other become an accepted social norm within the group. Only when the situation becomes unstable through death, emigration or invasion does there have to be an avert re-establishment of rank within a dominance hierarchy. So far as we have seen territorialism and dominance hierarchies as two types social systems which play a regulatory role in population dynamics, the first by regulating of space second by controlling individual priorities within a common space. In some animal species and populations, territorialism and dominance hierarchies apparently serve as sensitive feedback mechanisms which influence dispersal, mortality reproductive success and population size.

In other species and circumstances, territorialism and hierarchical structures are often subject to breakdown and collapse. Ecological consequences are then profound and extensive. In many rodent populations, excessive population growth and crowding is often associated with the following phenomena: a breakdown of stable territoria-lism and dominance structure, reduced fertility as evidence by lowered pregnancy rates, high prenatal loss and low infant survival, poor nest construction and poor maintenance, nest fouling and pollution, high aggression and wounding deteriorating health in the majority of the population, abnormal aggregations of individuals in the 'behavioural sink' pattern and finally population decline.

The detailed study to territorilism and dominance hierachies is really in the realm of ethology and animal behaviour, but the functions and consequences of these behavioural systems have a great deal of ecological significance. It has been abundantly demonstrated in many territorial animals that the holders of territory have the greatest breeding success. In fact, in some animals, non-territorial animals usually fail to breed. Similarly in those social groups characterized by dominance hierarchies the high ranking animals have the greatest breeding success and the low ranking members have less success or fail entirely in breeding efforts. In chicken flocks, low ranking roosters show frequent counting behaviour, sometimes more than dominant rooster, but they have relatively little success in completing copulations. It thus seems clear that the social systems of territorialism and dominance hierarchies have a limiting function in population dynamics; that is, they limit the breeding to those individuals which are ecologically and behaviorally most successful in establishing territory or high rank.

Theoretically, as population densities increase, and space becomes more limited, a smaller proportion of individuals within the population is able to achieve territories or high rank, and thus total reproduction is reduced in relation to population size. Ideally, these behaviors are density-dependent population limiting factors; that is, the higher the density, the greater the limiting effect on the population. The role of social behavior in population regulations has been discussed at great length by Wynne-Edwards. Wynne-Edwards proposed that the basic long-term population level of many animals is established by food and resource availability, but that social behaviour establishes a more immediate regulating influence on population size and density. This influence on social behaviour, he felt, comes into play before critical population levels are achieved; that is, before starvation and/or damaging violent aggression are conspicuous.

Thus, according to Wynne-Edwards most animals have intrinsic, self-regulatory population mechanisms which operate via social and behavioural feedback systems. These social and behavioral feedbacks are usually ritualized displays which relate to territorialism and dominance hierarchies, and thus convey information about population density. Therefore, regulatory feedback is provided according to the intensity of social competition. In other words, ritualized displays, or conventionalized rivalries, regulated breeding success and thus population numbers. Wynne-Edwards provides numerous examples of this hypothesis, including the complex display patterns of gallinaceous birds, (prairie chickens, sharp-tailed grouse, peacocks, bower birds et al.) waterfowl and many song birds. He also considers the vocal choruses of amphibians, the schooling of fish, and the flocking and birds and mammals as examples of complex social behaviour which serve the function of conveying information about population density and social competition.

Many aspects of Wynne-Edwards' theory are controversial especially when he theorized that most social behaviour has actually evolved as a means of controlling population levels. Wynne-Edwards proposed, for example, that territorialism has arisen as a social means of preventing over population. This theory has met substantial criticism from other ecologists and students of behaviour. These authors felt that there is little reliable evidence to support this hypothesis even though it is logically very attractive. Brown pointed out that, "Population regulation is never completely under the control of the species by itself, but depends in a complex way on interactions between members of the ecological community."

Brown felt one must consider the interaction of all factors, including predation, disease, food supply, interspecific dynamics of any given species. Brown found satisfactory evidence that territorialism tends to space individuals over available habitat, that it increases emigration and morality rates in individuals unsuccessful in finding territories, and that it has arisen in response to individual selection resulting from aggressive competition. The more detailed aspects of this controversy are out side the scope of introductory ecology. On the more basic parts of the hypotheses—namely, that social behaviour provides some regulatory feedback on population ecology—there is certainly no question that this is true in most animal populations.

Even this concession must not obscure the fact, however, that some animals have such poorly developed and/or variable responses to increased density and crowding that they can readily breed themselves into starvation or conditions of violent aggression and extreme social strife. White-tailed deer populations, for example have frequently increased to the point of extensive starvation. Lemming and other rodent populations have also increased to the point of mass behavioral strife where social mortality becomes extensive. Quite probably, some human populations are also in this category, providing example of populations in which social feedback limiting population growth often does not into operation until the point of starvation arrives (as in China and India), or until the point of violent and destruction aggression breaks forth (as in many wards in which population pressure has had a causative influence) or until social morality becomes limiting (as in countries where abortion is a significant form of population regulation). It is scientifically important to consider animal population in which environmental crises and/or behavioural breakdowns occur. In terms of human value judgments these seem to be cases of population regulation by wasteful and sometimes tragic means.

In evolutionary terms, however, many of the populations showing these phenomena are of highly exploitative natures, meet new environmentral situations had expand into available opportunities. Perhaps their lack of frigid, stereotyped behaviour and ecology, and their lack of precisely tuned population feedback systems, enables them to seize each new ecologic opportunity to the fullest extent.

Some of the best examples of such species are the domestic rodents especially the wild house mouse, the wild Norway rat (Pattus norvegious) and the Asian bandicoot rat (Bandicota bengalensis). Numerous ecological studies over the last 25 years have shown these species to have the following population characteristics:

1. They have tremendous reproductive potential and under certain environmental circumstances they often show invasive and eruptive population behaviour.

2. Populations can readily exceed the carrying capacity of the environment and this can result in sudden population decline.

3. Populations fluctuate widely, natural control is often abrupt and "ecologically wasteful", in terms of individual health and welfare.

4. Population regulation often involves social disruption and destructive behaviour.

These characteristics have been observed frequently in both natural and artificial populations of rodents over the past 40 years. The natural social behaviour of the Norway rat and wild house mouse includes both territorial and hierarchical behaviour. Females tend to the territorial around the nest site so that they normally exclude adult males and strange females. Often two of three females nest together in a communal agreement.

Males are also territorial if population density is not excessively high, and high ranking males defend a system of runway and the females living therein. The size and patterning of the territories depend upon the rank of the male, the density and social structure of the population and the nature of the habitat. In a stable habitat both population levels and social structures of domestic rats might remain quite stable over a period of time. But the habitats of domestic rodents are characteristically unstable with changing amounts of food (around farm, for example where harvests and food stores come and go), changing mortality patterns (as control compaigns) or disease outbreaks wax and wane) or changing cover conditions(as trash availability rises and falls) so that the populations are often experiencing drastically changing resource conditions. Under these circumstances, the populations respond rapidly with surges or declines of growth.

Under periods of rapid population growth leading to high densities, a variety of phenomena have been observed. There is often rapidly increasing social contact resulting in an increase in aggressive interactions. This frequently lead to a breakdown of territorial boundaries, increased crowding of the nest sites and sharply increased mortality of young nests become trampled and destroyed.

Females often desert the nests or actually kill the young. Wounds become frequent, especially an adult males forming sites for mites to invade the skin, so that acarine dermatits may achieve epidemic proportions. With a breakdown of territorial behaviour, bizarre aggregations developed so that clusters of individuals all in poor physical states appeared. Calhoun referred to such aggregations as "behavioral sinks". They are self-stimulating aggregations of individuals often collectively showing abnormal behaviour patterns.

Some individuals become inactive and socially withdrawn. As these conditions progress only limited portions of the environment are utilized but these become excessively polluted. Thus, the stage is set for both high social mortality and high disease mortality and population often collapses as rapidly as it built up. After the crash if the environment recovers the surviving individuals recuperate and population growth may begin again. The detailed of this theme differ considerably in different studies and different species. In come circumstances, there may be a cessation of production so that no young are born into the crowded population, as occurred in house mouse populations studied by Crowcroft and Row.

In other circumstances reproduction may continue, but there may be massive infant mortality of all young born. In still other cases, crowding may stimulate excessive migration and disposal as in lemmings and these mass migrations are often accompanied or determinated by high mortality. In all cases, a similar syndrome prevails. There is usually a drastic change in social organization and behaviour which most often have a major influence on reproduction and mortality. In domestic rodents these changes characteristically involve a breakdown in social stability and a disruption of normal territorial, reproductive and parental behaviour. Frequently parinatal processes are most dramatically affected, so that this pre-and postnatal survival of the young is reduced.

All population of living organism exist in a network of interactions with other population. Many of these interactions are subtle and complex than direct food relationships. Others are competitive or limiting to the interacting populations. The degree of intimacy in the relationships of the organism inhabiting a particular habitat varies greatly. These association may be either beneficial or harmful to both the patterns or beneficial to one partner and harmful to the other. If the relationship is beneficial either to one or both partners, it is called 'Positive Interaction' or cooperation. This may be interaspecific (Ex. Social life of inspects) or Inter specific (Ex: Commensalism, Mutualism & Symbiosis). The interrelations between individuals that are harmful to one or more of the participants is called Negative interaction. It includes "Parasitism, Predation and Competition".

The term 'Commensalism' was first proposed by Van-Benedan to the animals that share the same food. "Two or more animals that live together and do not enter into any kind of Physiological Union are referred to as `Commensals' and this association is called as `Commensalism! The relation is unequal in that only one individual gets the benefit from the association while the other is neither benefited nor injured. The Chief benefits from this association are (i) Shelter, (ii) Anchorage (iii) Transportation and (iv) The food supply.

A commensal which attaches or lives upon the host's body is called as 'ectocommensal'. If a commensal lives within the body of the host in the respiratory or alimentary tract or in any other body cavity which opens to outside, is referred to as endocommensal. Vorticella, a Protozoan animal is an ecto-commensal which is attached to the legs of the freshwater shrimp (prawn like) by slender stalks. The 'Sucker fish' is another ectocommensal which attaches itself to the body of shark, or sword fish or seaturtle with the help of its sucker.

All the ectocommensals are small in size and not burdensome to the host. The harmless protozoans in the intenstinal tract of mammals and micro-organisms in the canal system of sponges are the examples of an ectocommensals!. An example of commensalism is the remora-shark relationship where the remora fish (Echenesis naucrates) attaches to the skin of the shark by means of a strange sucker disc and transported widely and rapidly by the shark's motive power. The remora also consumes food remnants cast off from the jaws of the shark. Thus the remora benefits in two ways from this attachment and the shark is relatively unaffected although its swimming speed may be slightly impeded.

Many large organisms provide commensal harborage for other organisms externally and internally. Large tropical trees provide habits for numerous commensal plants and animals. Trunks and branches provide attachment sites for epiphytes such as orchids, and recesses between buttress roots provide attachment harborage for bats tree frogs, lizards, insects and many other organisms.

Whales provide attachment sites for barnacles, algae, and other sessile marine forms. These plants and animals are not parasitic, that is, they do not extract nutrients from the habitat provided by the host. Most animals, including man, also contain internal commensals. The human digestive tract is inhabited by several types of bacteria and protozoa such as Entamobea cole, which are neither parasitic not pathogenic, but merely residual in the alimentary canal. Another common type of commensalism occurs when various species of plants and animals use burrows or nests constructed by another species.

Termite nests provide ecologic niches for more than 100 species of other animals including ants, aphids, beetles millipedes and isopod crustaceans. Tube-dwelling annelids of the genus Chaetopterus which inhabit the tidal zone of the seashore provide habitats for small crabs of the genus Polyonyx. The crab benefits by the protection of the tube, and it apparently does no harm to the annelid worm which originally built the tube.

Commensalism may be obligatory, in which one organism depends entirely on another for its habitat or it may be facultative in which each organism may be independent but one is enhanced in the presence of the other. An example of obligatory commensalism is the relationship between the algae, Basicladia, and certain freshwater turtles. The algae grows only on the backs of the turtle. Similarly the bivalve mollusc Ostrea grows almost exclusively on the roots of the red mangrove off the coast of Florida.

Attached to the shell of the horse-shoe crab, Limulus, are several species of molluscs, barnacles and annelid worms which again are not parasitic but only residential and some of them live only on the Limulus shell. The oyster crab is a small crustacean which lives only in the mantle cavity is a small crustacean which lives only in the mantle cavity of oysters. These all represent cases in which one species of plant or animals has evolved a dependent relationship does not involve feeding upon the tissues of the host.

The possibility exists that this is one evolutionary route toward parasitism and the distinction between obligatory commensalism and parasitism is often difficult to draw. Facultative commensalism is represented by the relationship between the borrowing owl (Speotyto cunicularío) and the prairie dog (Cynomys sp.) The burrowing owl often nests in prairie dog burrows but is not confined to such burrows. In the north polar regions, the arctic fox Clo pex lagopus) feeds in the winter time on the remains of seals killed by polar bears (Thalarctos maritimus) though it may have other sources of food as well.

Probably the majority of commensal relationship are facultative. It is especially interesting to consider animal populations which form facultative commensal relationships with men. Among vertebrates perhaps the best examples are domestic rodents such as house mice and Norway rats which have increased in abundance and distribution by their association with human habitation and agriculture. The house sparrow, pigeon and Asian house crow and Asian house shrew or mūsk shrew have also greatly increased their population sizes through human associations.

Cities, towns and farms provide increased food supplies and nesting sites of these animals. The rhesus monkey in India is one of the few primates which has become a facultative commensal of man. The rhesus can readily survive in forests without man, but it becomes especially abundant around villages and towns of the Gangetic basin where it could live on agricultural produce. Some more examples among the inverterbrates are as follows:-

(1) The peacrabs are found living in the mouth cavity of bivalves. They steal food collected by the host and do not cause any injury.

(2) A number of micro organism live as commensals in the canal system of sponges.

(3) Polynoid polychaetes are found in the pallial chambers of gastropods and in the ambulacral groves of star fishes.

(4) A host of commensals such as worms, Brittlestars, molluscs, little fishes, prawns-crabs etc. are found among the coral reefs.

(5) Small pelagic fishes live under the umbrellar shelter of large jelly fishes.

(6) Some fishes live in the borrows of worms and crustaceans.

(7) A number of copepeds live within the Ascidiann or Sea anemones.

(8) Barnacles attach on the skin of whales or back of turtles.

In all these associations one of the partners which is generally smaller derives benefits of anchorage, shelter, food and transportation.

Mutualism is an association between two or more animals in which all derive mutual benefit with no harm to any of the partners. "Symbiosis" term also often used to this type of relationship (Properly refers to the intimate associations to two or more dissimilar organisms). Mutualism may be seen between animals and plants and between animals and other animals.

Ist Type : Mutualism between Plants and Animals

This association is important to the reproduction of many kinds of plant species. Plants attract insects by their sweet, nectar or colour. The insects visit plants to get their food. The plants dust them with pollen and the insects carry the pollen from flower to flower.

In the distribution of seeds, plants take the help of animals. Plants provide fruits and seeds pass in their hard coat, undigested through the alimentary canal of the animal and are distributed in its droppings. For instance a species of Alacia grows little pearshaped masses of food for the ants, and hollow thrones (Myrmecolphilous stipules) in which the ants take shelter. In turn the ants defend the plant from the attack of other insects.

2nd Type: Mutualism between Animals and other Animals

Birds display this type of relationship. The cow bird and Bison, the Oxpecker and the antelope, crow and the cattle exhibit mutualistic relationship. In all these associations the birds render considerable service to their hosts by ridding them of ticks and external parasites and also serves as sentinels. In turn the birds receive a constant food supply.

An interesting example is between a 'Crocodile' and the bird (Pluvianus aegyptius). The bird removes leeches and fragments of food in decaying condition of the crocodile. The crocodile opens its jaws and permits the search then finally the bird receives the food from the crocodile. The common starling bird may also perform the same function from the sheep. The association of ostriches and Zaebras is also said to be of mutual benefit.

Birds and Insects

Birds have also developed relationship with insects for mutual benefit. The horn bill of Africa harbours a number of insects in its nest. The female horn bill will be confined to the nest at the time of egg laying and breeding. The male bird feeds her with fruits.

Various ants plants (Lice Aphids) are associated as Commensals. The most common Commensals among marine forms are the Hydroids and Sea anemones found on the shell of Hermit crab. In this association both of them are benefited. The Indian Rock (Mincus inermic services as a moving home to zoophytes small polyps are encrusted on this fish. In this relationship the fishes obtain protection from predators and they pay for it by luring large fishes within the range of tentacles.

The relationship between *Hermit Crab* and *Sea anemone* is very interesting. Here the Crab obtains protection from predatory fish by the shrinking power of sea anemone. The sea anemone also gets an advantage of transportation of feeding grounds and gets food fragments when the crab captures and eats another animal.

Sponges growing on the shell of Hermit Crab protect it by producing bad odour. Here the advantages to the sponge is that it is carried to new waters for its own food. Thus, the mutualistic relationships between plants and animals, and animals and other animals prove to be beneficial mutually and reciprocally to the participants.

Like commensalism, mutualism may be obligatory or faculative. An example of obligatory mutualism is provided by the relationship between the flagellate protozoan Trichonympha, and woodeating termites. These species cannot live without each other. The protozoan lives only in the digestive tract of the termite. The stomach and caecum of the horse contains millions of ciliate protozoa and bacteria which digest cellulose for the horse and provide 20 per cent of its nitrogen requirement per day. These intestinal microorganisms are essential for the normal growth and health of the horse.

An elaborate form of obligatory mutualism occurs between some ants (e.g., Cremastogaster lineolata) and aphids or plant lice (e.g. Aphis caliginosa). The ants maintain the aphids in specially constructed nests, and they feed upon a nutrient liquid secreted by the aphids.

The aphids have sometimes been called "cows" because they release the secretion when stroked by the antennae of the ants. Thus the ants receive food from ants. Many other anthropods, including mites, springtails, flies, wasps and beetles, have developed similar complex relationships are based on interspecific communication systems through which the different species jointly utilize food, shelter of other resources. Many mutualistic relationships are facultative; that is, both populations can exist without the other, but both are favoured when living together.

Baboons and impalas, for example, can survive well without each other, but each provides a unique protective warning system for the other when they live in association. Squirrels facilitate the extension and propagation of hickory trees by buying the nuts, but the hickory trees can propogate without squirrels and the squirrels can likewise survive without hickory nuts, if an adequate supply of other food is available.

Man has capitalized on mutualism as a biological principle in the development of agriculture. The domestication of plants and animals originated in facultative commensalism or mutualism. At the down of agriculture, 10,000 years ago, man could live without domesticated plants and animals, but as he developed these relationships then evolved into obligatory commensalism and mutualism. Modern man obviously cannot live without the domesticated plants and animals of agriculture, and many of these domesticated forms cannot live without the special husbandry provided by man.

The term Symbiosis was first coined by DeBray in 1879. It means "living together." "Symbiosis is an association between two or more organisms that live together in close physiological union for the mutual benefit." Here the term Symbiosis refers to the intimate association of two or more dissimilar organism. One of the partners may be known as Symbiont. Symbiosis offers from commensalism and mutualism in two important respects.

(1) The relation between two partner is much more intimate in Symbiosis than in other relations.

(2) It is mostly obligatory.

Symbiotic associations are known to occur between plants and other plants, animals and plants; and animals and other organisms.

First Type Symbiotic Association among Plants

This type of associations is demonstrated in (i) Fungi and algue which form lichens; (ii) Fungal mycorrhizae and the roots of many flowering plants and (iii) Nitrogen fixing bacteria and the roots of legumes. In all these associations, there is metabolic interdependence. In the lichens, algae produces food and fungi provide mineral and protection.

The intimate association between the roots of flowering plants and mycorrhizae results in the formation of a composite fungus-roots structure which differs from the roots without mycorrhizae. Here the fungi require soluble carbohydrates and substances and they get them from root secretions. Here also the association is intimate the benefits are mutual. The importance of the Nitrogen-fixing bacteria (Rhizobia associated with legumes is well known.

Legumes posses on their rootlets, a number of swellings, inhabited by the bacteria. Here, they derive carbohydrates from the host and in turn they fix the atmospheric nitrogen in the roots. This "Nitrogen" is used by the plants. This is clearly a beneficial partnership.

Second Type: Symbiotic-Associations Between Plants and Animals

The most notable symbiotic association between plants and animals occur between freshwater coelenterate (Chlorohydra Viridissina) and an algae. In this association the animal obtains Carbohydrates and a supply of Oxygen evolved as a result of Photosynthesis by the algae. The large in turn receive protection and Dioxide for photosynthesis released by the animal during respiration. Hydra and Algae relationship is also one of the good examples. The mutual benefits of hydra an algae are

Chlorachydra (*animal*)	*Zoochlorellae* (*Plant*)
i) It gives phosphorus and nitrogen compounds for protein formation.	i) Plants gives Oxygen
ii) It gives Carbon dioxide for the photosynthesis in plants.	ii) Excretory materials.
iii) It gives protection to algae (Plant)	iii) Carbon dioxide (from respiration)

Planeria and an animal algae and a plant symbiotic association is also a good example. Here the symbiotic benefits are same as in coelenterate and algae. Here, the presence of algae is indispensable for the worm and this is obligatory.

The wood boring beetles have an elaborate apparatus on the head for carrying the Fungus. Others carry Fungus in the midgust to be regurgitated for cultivation. The females of this species have an apparatus connected with evipositor which coat their eggs with fungal spores as they are laid. In all these associations, the Fungus provide food and shelter, and it grows on the excretory material of its host.

Third Type: Symbiotic Association Between Animals (Metazoa) and Microorganisms

Symbiotic associations are also seen between Metazoa and Micro-organisms many Wood-eating-insects especially termites harbour numerous protozoans in their alimentary canal. It has been shown that the gut of termites can live on a nitrogen-free diet so long as protozoa are present. This suggests that they may be capable of nitrogen fixation of the synthesis of protein from carbohydrates. Here the protozoans produce specific enzymes "Cellulose," and "Cellobiase," which are responsible for cellulose digestion. Similar association is seen in the gut of Cockroach, "Cryptocersus." In all these cases the protozoan gets shelter and food.

Mammals and Bacteria

In this association symbolic bacteria are present in the Caecum of rabbits and rumen of cattle. Here, the bacteria utilize the inorganic nitrogen and sulphates for the synthesis of essential amino acids, which can't be synthesized by the animal. In turn to mammals, the Bacteria produce Vitamin B12 (in ruminants). In man also some of the vitamins are produced by the intestinal Bacteria. The Symbolic Bacteria are also present in the gut of the South African Honey Guide." Here, the Bacteria feeds on the Wax of honey comb the bacteria digest the Wax and gets shelter and food from the host.

Predation

In this relationship one animal is killed and eaten by another, the predator so predator is a free living animal. It is usually larger than prey. The herbivores that feed on the plant material completely are called as Plant Predators. There are, many kinds of predation which are not necessarily so selective. The swallow feeding on moths and midges, the top minnow or trout feeding on mosquitoes, and the ant eater feeding on an ant nest is less discriminate in taking individuals.

Many able and healthy individuals are preyed upon. In cases such as these, which involve large numbers, the prey population often shows great productivity which can accommodate this mass mortality. Many fish, insects, and other Invertebrates have tremendous production of eggs and larvae, but only a small percentage of these serve the onslaughts of predation and other forms of loss. For example, the horse shoe crab or Limulus of estuarine and evastal waters deposits its eggs on sandy beaches wehere they are consumed in great numbers by flocks of laughing gulls and other shore-birds. Large number of eggs also wash out to sea and are eaten by fish, whereas others are pushed up on the beach and dry out.

Despite this fantastic loss of eggs and young larvae, Limulus continues to survive in large populations and has done so for the last 180 million years. It is one of the oldest animals on earth in terms of its evolutionary history, having outlived trilobites and dinosaurs by millions of years, and it is still incredibly abundant. There has been considerable interest in predation as a factor in quantitative population control as well as its role in quality control. On one hand, sportsmen have wanted to limit predation, being concerned that predators may take excessive numbers of game birds and wildlife. Some sportsmen's organizations have therefore supported predator control campaigns against hawks, owls, foxes, coyotes, wolves, cougar and other predators in the name of conservation.

On the other hand, ecologists have often been interested in promoting predation as a method of biological control of pest animals, particularly insects and rodents, and even as a necessary biological control of desirable animals such as deer, or prevent them from becoming pests. There are also examples from the entomological literature where predation serves as a significant population control on an animal population. In a classic study of natural population balance in the knapweed galifly (Urophora jaceana), a small dipteran insect, Varley concluded that one of the major controlling factors was the predation of mince upon insect pupae. His data showed that 22 to 43 percent of the fly morality was attributable to this predation.

One study of tsetse flies (Glossing spp.) in Africa, found predatory spiders to be a controlling factor on the abundance and distribution of the flies. Insect predators have also been used in applied biological control of agricultural pests in a few cases. In California, the cottony cushion scale, an hemiptoran insect pest of citrus crops, was effectively controlled by a predatory ladybird bettle imported from Australia. In Hawaii, the sugarcane leaf hopper has been controlled by the capsid bug.

Another important aspect of predation as an influence in population ecology and natural selection is the great variety of adaptations which have arisen in response to it. Protective coloration, warning coloration and mimicry fall into this category. Also, the morphological and behaviour, freezing behaviour and many other specialized patterns have developed in various species as adaptations to reduce mortality through predation. One particularly interesting system which has been studied in some detail is the intimate relationship between certain moths and the bats which prey upon them.

The bats locate the moths by emitting ultrasonic pulses and detecting reflecting echoes flying moths, a natural sonar system. The moths have evolved the ability to detect these ultrasonic pulses of the bats, and upon perceiving the approach of a feeding bat, they instantly go into complicatd diversionary flight spirals to avoid the bat. The introduction becomes on serial "dog flight" between bat and moth, with the moth trying to avoid the bat. This ability and behaviour has evolved in selective response to predation in the same way that cryptic coloration has evolved. In both case the advantageous mutations have been selected and propagated.

"Parasitism" is the relation between two individuals, wherein the parasite receives benefit at the expense of the host, therefore, parasitism is form of disoperation. In this association, the parasite receives (i) food (ii) shelter (iii) and protection, at the expense of the host. The host does not derive any benefit from the relationship. A parasite does not ordinarily kill its host atleast not until the parasite completes its reproductive cycle. Parasites are classified into different categories depending upon their relation with the host.

1. *Temporary Parasites*

The animals which spend only a part of their life cycle as parasites are called temporary parasites."

For Example; The glochidium leads a parasitic life. It attaches itself the skin of fish with the help of hooks present on the shell and later it is covered by the skin of fish. It remains burried thus for several weeks and emerges as young animals to lead a free life.

In prasitic Hymenoptera, the eggs are deposited into the body of another insect (Caterpillar), where the entire larvae stage is spent the larva feeds on the fat body of the host. Until it attains pupal stage. Finally it emerges as an adult and leads a free living life.

2. *Permanent Parasites*

Some parasites have continuous association with their

hosts, they are called permanent parasites. They spend their entire life as parasites.

For example: Trichinea worm, Malarial Parasite etc., Permanent prarasites often found among lower animals. They almost always require a succession of hosts or intermediat host.

3. *Facultative Parasites*

These are parasites (animals) which are parasitic only when there is a change for parasitic life, leading free life otherwise. Eg: Meggocs of some flies, Lamprey etc.,

4. *Obligatory Parasite*

Some species are forced to lead a Parasitic life. If not, they can't live. This type of parasites are called obligatory parasites. Eg: Filarial Wornm Cestodes etc.,

5. *External Parasites*

Those which live outside the body of the host are called external parasites. Eg. Flea on birds and mammals, blood sucking lice and fleas, mites, ticks and leeches.

6. *Endoparasites*

These are parasites which live in the elementary tract body cavities, various organs, blood or other tissue of the host. The association between 'endoparasites' and their hosts is invariably more inntimate than that between extoparasites and hosts. Tricchomonad flagellates, Opalnid ciliates Sporozoans, Nematodes, Cestodes and some Copepods are some of the endoparasitic forms.

The endo-parasites are further divided into 2 types based upon their exact location inside the most body.

(i) Intracellular Parasites Ex:- Plasmodium (living within the cell)

(ii) Intercellular Parasites Ex:- Tryponasome (living be tween the cells in the intercellualar spaces).

The origin of parasitic life in amimals is associated with attempts to secure food for themselves or their offspring. The ancestors of ecto-parasites were clearly free living forms. A free living organism could accidentally have settled on the body of a larger species where coditions were favourable for survival. For example: The biting lice might have evolved from the insects that live beneath the bark of trees. Endoparasites might have evolved from ectoparasites or directly from free-living ancestors.

Modifications and Adaptations

1. *Organs of attachment:* The rasping mouth of the Lampern, and the head of tape worm equipped with 'Suckers' and 'Hooks' for attachment to the hosts gutwall are the typical examples. Among parasitic arthropods such as mites other insects claws are well developed used to cling on the external surface of the host.

2. *Body Shape:* The intestinal parasites such as the tape worm and Round worm have elongated bodies. The liver fluke have leaf-like bodies and elongated form of blood flukes are good example of this type. These arrangements are useful to suit the environment in which the parasite lives.

3. *Cuticles:* Ex: Tape worm. Intestinal parasites face the problem of with-standing the action of the digestive juices of the host. To avoid this these parasities develop thick cuticle, offers resistance against the action of the digestive juices.

4. *Loss of Locomotary Organs:*Ex: The fleas or Lice. Most of the endoparasites have lost their Locomotary organs, as they are not needed. There is no need in search of food, as they have always plentiful supply of food. So the locomotary organs of the parasites disappear or differentiate into organs of attachment. The fleas or lice (Parasitic Insects) have lost their wings, which live among the hair and feathers.

5. Digestive System: In the tape worm, which lives in the alimentary canal of the hot, has lost its alimentary canal and absorbs already digested food from the surrounding medium. The mouth parts of the blood sucking insects have developed 'Piercing and Sucking' apparatus.

6. Loss of Sense Organs: As the internal environment of the host is more or less uniform, there is no use of the sense organs among the endoparasites are provided only with tango receptors.

7. Complicated Reproduction: Ex: Flat Worm, Round Worm, etc. The reproductive function in most parasites has undergone considerable complication. It is the goal of all the functions of the organism. The endoparasites are characterized by enormous fertility. For example (i) the human round worm lay (Ascaris Lumbricoides) eggs at a rate of 1500 a day (ii) the termatodes and cestodes lay eggs continuously (iii) Taenia solium (Tape Worm) lay eggs annually and have been estimated at 80 million (per year).

The most remarkable feature of flat Worm parasities is the presence of 'complicated reproduction'. The effect of "Parasitism" on the host is usually harmful. The endoprasites causes effects, more marked than those by extoparasites. The following Pathological effects caused by parasites in their hosts.

(1) The destruction of cells and tissues of the host by the parasite's movement or feeding act.

Ex: Entamoeba histolytica-eats the tissue cells of colon and red blood cells of the host, and plasmodium feeds on liver cells and erythrocytes (R.B.C.)

(2) Parasites may cause 'enlargements's and disorders of lymph glands, and liver.

Ex: Leishmania.

(3) Parasites may cause 'Malaria', (Plasmodium parasite) and 'Trypanosoma gambiense' diseases due to the secretion of toxins (Poisonous).

(4) The effects of endparasites may also be mechanical. Ex: Ascaris, is a good example when these parasites multiply in greater number they block the intestine, preventing the feree flow of food, which leads to the death of the host.

(5) The most remarkable effects of prasitism on the sexual characters are seen among crustaceans. The presence of sacculina bring spectacular changes in its host (Crab), which may result finally in death (due to the parasitic castration).

Sometimes a parasite harbors a number of parasites in its body. They are called 'Hyperparasites'. A typical example is, small mammal ones as 'Mouse' which invariably harbours a no of ectoparasitic fleas, which themselves contain endoparasites (Protozoans) in their alimentary Canal.

The parasite which do not cause any injury to the host are called 'Non parthenogenic parasite'. The parasite which cause disease is known as "Parthenogenic Parasite". In man's digestive system 'Bacteria' are helpful in synthesizing' Vitamins, used to man. Some Bacteria causes numerous diseases as pneumonia, tuberculosis, diptheria, bubonic plague, cholera, syphilis and wooping cough etc.

Viruses causes, small pox, polio, yellow fever etc. Animal-like microbes causes malaria, (Plasmodium) sleeeping sickness (Trypanosona gambiense) and dysentery (Entamoeba histolytica). Parasitism is virtually universal in all plants and animals in vertebrate animals, endoparasites are found within many of the major organ systems of the body, most commonly in the digestive, circulatory, respiratory and urogenitial systems. Ectoparasites occur on or within the skin and its appendages, such as hair scales. Throughout most of the world, man is beset with many parasitic organisms including intestinal worms (tape-worms, flukes, roundworms, etc.), intestinal protozoa (amoeba, ciliates and flagelates), parasites of the blood (microfilaria, malaria protozoa, etc.), and a variety of extoparasites (Mice, mites, ticks, mosquitoes, etc.).

Many infectious microorganisms including bacteria and viruses are also parasites in the sense that they extract nutrients from the host. Some of these are pathogenic, that is, they impart the health and normal functioning of the host, as for example, the bacterial organisms of the genus shigella which cause bacillary dysentery. Others, such as some intestinal flagellates, are nonpathogenic and do not harm the host.

Many parasitic organisms may be pathogenic in one individual and nonpathogenic in another. For example, most individuals, infected with Entamoeba histolytica in the digestive tract do not have amoebic dysentery, and, in fact, may have no disease symptoms at all, but in other individuals, the same organism may produce severe disease and even death. This is a matter of individual ecology and immunology which is not well understood. As mentioned previously in the selection on communalism, theere is no sharp line of distinction between obligatory commensalism and parasitism.

Theoretically, Entamoeba histolytica feeding on tissues of the host is a parasite, whereas the flagellate Trichomores hominis, feeding on digested foods before incorporation into host tissue is not a parasite, nor is Entamoeba coli feeding on undigested particles of food and intestinal bacterias. Neither is there a sharp line of distinction between a parasite and a predator. Houseflies and vampire bats bite and suck blood form livestock, and they might be considered either parasites or predators. Normally, we think of predators as killing their prey in a short time. Man's relation to the cow is predatory when he kills it for beef, but it is parasitic when he milks it for dairy products.

There are some relationships, however, in which the distinction is arbitrary, as for example, in the lamprey feeding on a host fish. The lamprey may kill the fish in a few days or a few weeks, depending upon its size in relation to the fish, and may be considered either a predator or a prasite.

Ecologically, one of the most interesting questions about parasitism is the effect of the parasite population upon the host population. The oldest and best adapted parasites have little or no pathogenic effect upon the host. Pinworm infections in children are relatively harmless, though they may cause a minor irritation around the anus. In West Bengal, studies have shown that over 75 percent of the local population harbors hookworm (Necator sp. and Ancylostoma sp.), but there is no demonstrable pathology or effect on the people. The infectious burden (i.e., the number of parasites per person) seems to be small and insignificant.

Most wild animals have parasites are able to maintian excellent health with these parasites. Zebras in Kenya are intensively infected with many internal and external parasites. but they are usually in remarkably robust health. On the other hand, there are numerous examples where parasitic infections are harmful and debiliting to the host. Certainly, hookworm necator americana in the southern United States, as it occurred 50 years ago, was a debilitating disease, sapping the strength and health of the infected person. Throughout the world, malaria still remains as a major health problem and it causes a great amount of illness and human misery.

One of the most important debilitating disease of modern times is schistosomiasis an infection of blood flukes of the genus Shistosoma. This is a waterborne infection and has spread throughout the tropical world with the expansion of irrigated agriculture. In animal populations, parasites may also weaken infected individuals. In the bighorn sheep of Wyoming and Idaho, lungworm infections are still a major cause of illness and mortality. In dogs and wolves, heartworms reduce the animal's vitality and hunting succss. In meadow vole populations, botfly parasitism causes larger spleen weights in infected mice and reduces their survival when exposed to cold temperatures.

Periodic reductions of field voles in Wales have attributed in some years to heavy infections with tuberculosis among the mice, but in other years similar reductions occurred without the infection. Similarly, reductions in populations of red grouse in Scotland have been attributed in some years to the round worm parasite, Trichastrongylosus pergracilis, but this has not a consistent finding in all year. Occasional heavy mortality of gray squirrels in Baltimore has been associated with coccidiosis, caused by a protozoan parasite of the digestive tract. In 1971 a virus disease of wildfowl decimated 90 percent of the pheasants of England, in much the same way as myxomatosis reduced rabbit populations of England, and Europe in the 1950's. Thus, parasitism as an interspecific population relationship may be balanced or unbalanced in regard to its effect upon the host.

A seriously unbalanced relationship, in which the host dies is, of course, selectively disadvantageous to the parasite as well. Another outstanding feature of parasitism is the interspecific complexity which was evolved in some species. Many parasites have primary and secondary hosts and, in some cases, tertiary hosts for various stages of the life cycle.

The malaria organism Plasmodium falciparum, for example has parts of its life cycle, the sexual phase of reproduction, within the mosquito, and part, the sexual phase of reproduction within the human host. Schistosomes infect man or other vertebrate animals as the primary host for sexual reproduction. The Chinese liver fluke, Clonorchis sinensis, infects man as a primary host, aquatic snails as a secondary host, and fresh-water fish as tertiary hosts. Despite the complexity of this life cycle, it is a very successful and widespread parasite, infecting millions of people throughout the Orient.

The word "competition" denotes striving for the same thing. At the ecological level competition becomes important when two organisms strive for something that is not in adequate supply for both of them. Thus, plants compete for light and nutrients in a forest, and animals compete for food and shelter when the latter are relatively scarce in terms of the density of animals. If the population consists of only a few scattered individuals, competition will not be a factor of ecological importance. In the arctic, for example, plants may be so few and scattered due to the severe climate that no competition for light occurs. The result of competition is that both parties (that is, the competition) are hampered in some manner. At the population level this means that density or rate of population energy flow, will be reduced or held in check by the competitive action.

There is another type occurs when two organisms interfere with one another while striving for something even though that something is not in short supply. For example, the organisms might secrete substances that interfere with each other, or they might even eat each other. Many ecologists prefer not to include such direct mutual inhibition under the heading of competition. Both intraspecific and interspecfic competition can be very important in determining the kinds and numbers of organisms. Intra-specific competition is an important factor in those populations that tend to be selfregulated in the manner described above.

An interesting behaviour pattern, which results in intraspecific competition for space and a rather effective control of population size, is known as territoriality; it is characteristic of many species of birds and some other higher animals. At the beginning of the breeding season the male of a territorial species of bird will "stake out" a definite area of its habitat (that is, the "territory") and defend it against other males of the same species, with the result that no other ale is allowed to enter the area. Much of the loud bird song we hear in the spring is for the purpose of announcing to other males "ownership" of land, and not for the purpose of wooing the female, as is often supposed.

A male that is successful in holding his land has a high probability of mating and nesting, while a male that is unable to establish such a territory will not breed. Once the pair is formed, the female also joins in the defense of the territory in many species. The defended territory also serves a positive purpose of insuring the complex business of caring for the nest and young will not be interrupted by the presence of other males and females. Finally, it should be mentioned that the actual defense of the territory does not usually involve much actual fighting or other severe stress.

Would be invaders respect the established bird; loud songs or threat displays usually are sufficient to cause the invader to withdraw. However, if one of the pair should be killed it is very often quickly replaced by a bird form a reservoir of individuals not established. Therefore, since the territorial behaviour pattern helps avoid both overcrowding and undercrowding it can be siad to be regulatory in the sense that such behaviour promotes a sigmoid growth pattern, or at least helps prevent an overshoot of carrying capacity. Plant populations, as well as animal populations, may regulate themselves, so to speak, and avoid overcrowding.

As one drives through the desert in the southwestern United States one is impressed with the fact that desert shrubs are widely spaced, often almost uniformly distributed as if planted at regularly spaced intervals. The pattern would seem to be explained by competition for scarce water, which eliminates all but one individual in a given area. However, evolution of self-regulation of the population has occurred in some species, in that severe competition for water is avoided by the production of leaf or root hormones that inhibit development of other individuals in the neighborhood.

Chemical substances rleased by decaying leaves that fall to the ground under the shrub, or by the living root in the soil, inhibit or kill any seedlings that may start to sprout. Such a control mechanism would tend to keep plants spaced apart, thereby reducing actual competition for water, which might result in stunting or death of all plants should there be many trying to grow in the same place. In this kind of "birth control" the quality of the individual in a limited environment is maximized, not the quantity; is there as lesson here for man?

Amensalism is an interspecific relationship in which one population is inhibited while the other is unaffected. A simple example is the shading out of certain plants under tall trees. The trees reduce the available sunshine at the ground level, and numerous species of plants cannot find adequate light in the shade. Hence, only shade-tolerant plants with lower light requirements can survive as ground cover in the forest. Antibiosis is a specific type of amensalism in which one organism produces a metabolite that is toxic to other organisms. The best known example is the mold, Penicillium, which produces an antibiotic substances causing the death of many bacteria. This was observed and identified in the classic work of Florey and Fleming in England in the late 1920's and early 1980's which led to the development of antibiotics in clinical medicine.

Penicillin, strepotomycin, aureomycin and other antibiotics used against disease organisms all represent a similar ecologic principle, and one which is widespread in nature. Many species of fungi and lichens produce metabolic substances which inhibit bacterial growth. The green algae Chlorella produces substances which inhibit the growth of diatoms of the genus Nitzschia, and conversely, Nitzschia produces a substance which inhibits Chlorella. Many plants also produce substances which are toxic or inhibitory to animals. Old cells of Chlorella produce a substance which inhibits the feeding of Daphnia. Algal blooms of some blue-green and red algae produce chemicals toxic to fish.

The well known Red tide of southern coastal waters, produced by the flagellate Gymnodinium brevis, can result in massive fish kills. Land plants also produce inhibitory substances. The roots of some species of trees produce substances which inhibits the growth of other trees, both of their own species and different species. This may result in a characteristic spacing pattern of trees within a forest, dependent not so much upon competition for sunlight at the conopy level, but upon competition for space, water or nutrients at the subterranean root level. Many plants produce substances toxic to animals which touch them or eat them. In Nevada and Utah the desert bush Halogeton glameratus produces oxalic acid which is poisonous to sheep. Poisonous toadstools to and tubers have been well known throughout the history of man. In South America and Africa, many native people harvest the tubers of the poisonous manioc plant, but they place it in a series of soaking vats to leach out the toxic chemicals before consuming the starchy tissue of the roots.

A small African monkey, the talapoin monkey (Cercopithecus talapoin), has learned to steal the manioc from the soaking vats at just the proper state. Many of these substances are in a class of compounds known as ectorines of environmental harmones. Ectorines are substances produced by the organism which affect other organisms of the same or different species. Not all ectocrines are toxic or inhibitory, of course; many are stimulatory or beneficial. But collectively, they represent chemical messengers by which individuals and species are interrelated.

All organisms react on their habitat in one way or another, and when they occur in numbers these reactions produce a conspicuous effect. Water conditioning occurs hence physical or chemical changes occur as the result of organisms living in it. Compared with unconditioned water, these changes may have either a harmful or a beneficial effect on organisms introduced into the water after the original organisms have been removed. Water is said to be homotypically conditioned when the changes were previously produced by individuals of the same species as being studied and heterotypically conditioned when the changes were produced by a different species.

Experimental studies have demonstrated that goldfish grow faster in water that has been homotypicaly conditioned for 24 hours than in unconditioned water. Both fish and amphibian larvae also do better in water conditioned by the presence of mollusknes than in unconditioned water. The marine flatworm Procerodes wheatland will survive much longer when transferred to fresh water conditioned by the presence of either live or dead individuals of the same species or by freshwater species of flatworms than they do in unconditioned fresh water.

The longer survival in toxic solutions, faster growth, and greater reproduction of protozoans, snails, flatworms, cladcerans, amphibian larvae, and fish occurring in aggregations rather than as isolated individuals is attributable to water conditioning. Benefits derived from aggregating are shown in other ways by terrestrial animals. In honeybees, when hive temperatures drop below 14°C (57°F) during the winter, they form clusters and maintain a mass temperature several degrees above outside temperatures. This is brought about by increased metabolic oxidation of honey in their bodies and by increased muscular activity.

Furthermore, the compact cluster presents a surface area for heat loss that is less than the total surface area of the individuals separately. When there is danger of overheating, the bees in the hive spread out on the combs and fan with their wings to create a circulation of air. They will also carry water into the hive and place small quantities both outside and inside the comb cells. The forced air circulation evaporates the water and cools the hive. Bees also cool themselves by constantly moving their tongues in and out of their mouths, exposing to evaporation the moisture that is present on them as a thin film temperature regulation is less well developed in other social Hymenoptera.

Included here are the coactions whereby chemicals secreted by one organism affects the growth, health, or behaviour of other organisms. Allelopathy is produced in plants when toxins are liberated that inhibit seedling growth in the vicinity. This may affect succession of plant species, especially important in the early stages. Some pioneer species in the abandoned field sere produce substances inhibitory to nitorgen-fixing and nitrifying bacteria. This retards invasion of other species that require higher nitrogen concentration in the soil.

Volatile inhibitors are generally more prevalent than water-soluble ones, and relatively more prevalent in arid than humid climates. Antibiotics produced by bacteria, fungi, actinomycetes, and lichens are widespread in nature and may be one of the reasons why bacteria pathogenic to man cannot mutiply well in soils. A number of antibiotics, such as penicillin, have been used extensively in human medicine. The use of allelochemic coactions agriculture has possibilities but has yet to be exploited. Among animals, overcrowding of tadpoles in culture dishes in associated with the occurrence of peculiar round vacuolated cells in the intestinal tract and feces that appear responsible for curtailment of further growth. Under laboratory conditions, killer stocks of Paramecium aurelia produce a toxin, paramecin, at the rate of one unit-particle per animals per 5 hours. One unit-particle is enough to kill one individual of sensitive stock of the same species as well as being lethal to other species of Paramecium.

Conditioning that becomes unfavourable homotypically may sometimes be favorable, or at least tolerable, heterotypically. Thus in protozoan infusions there is a microsere of one species succeeding another. Allechemic effects are of great variety in both plants and animals: repellants, escape substances, suppressants, venoms, inductants, attractants, signals, stimulants, autotoxins, antoinhibitors, and so on. Pheromones chemical messages between members of a species especially important in reproductive behaviour, social regulation and recognition, alarm and defense, territory and trail marking, food location, as so on many of these effects are beneficial to the individual; others serve for competitive purposes.

In the positive relationship between populations, proto-cooperations is a short step ahead commensalism, to cooperation, both populations gain by the association or interaction of some kind. A considerable work on protocooperation was one by Alee. In natúre, crabs and coelenterates are often found associated with each other. This association is of mutual benefit to both. The coelenterates grow on the back of the crabs, or sometimes planted there by the crabs. The coelenterates provide protection to the crabs. In turn, these are transported by crabs to new situations, which ensure them good supply of food. There is no physiological relationship betweeen the coelenterate and the crab, and none is absolutely dependent upon the other.

3 Population Growth

Population studies are fundamental to many human problems in agriculture, pest control, wildlife management, and economic planning. The purpose of population research is to understand the interplay of factors influencing population range, and ultimately to predict and regulate the course of populations. Man obviously wants to limit or reduce populations of organisms, whereas he has wish to maximize within reasonable limits valuable and productive animals for aesthetic or economic reasons. The greatest challenge of all, as well as scientifically and philosophically the most difficult are of human endeavor is understanding and regulation human population change.

There is no doubt in the fact that human population are drastically out of balance. For thousands of years prior to the seventeenth century, the population of the world was less than 500 million people. Human population were limited by high mortality and even in the great civilizations of Egypt, Greece and Rome, expectation of life probably did not exceed 30 years. The class checks on human population were disease, famine and war. In the fourteenth century, at least 25 percent of the adult population of Europe died in epidemics of bubonic plague and between the years 1348 and 1379 A.D. England's population was reduced almost 50 percent, Cholera, typhus, small pox, malaria yellow fever and sleeping sickness have all been among the great killers of man throughout history. Several of these diseases, especially malaria and yellow fever, severely limited the abundance and distribution of people in tropical countries until the advent of modern medicine and public health. Asia has traditionally been the great stage for famine though starvation; death also occurred in Europe. The famous Irish famine of 1850 resulted in the death of more than 12 percent of the adult population between 1846 and 1851. In China, in 2000 years preceding the 20th Century, over 1800 separate famines were recorded, many of them killing million of individuals. More recently, 4 million famine deaths occurred in China in 1920-21 more than 5 million is Russia from 1918 to 1923; and over 2 million in India in 1943. Almost agriculture and transportation have alleviated famine mortality, malnutrition continue to be significant demographic for, particularly since it accentuates illness and mortality from infectious disease.

War has also been a major mortality factor throughout history,not only by causing direct battle casualties, but also by stimulating disease and famine. Great epidemics and massive crop failures, have often been triggered by the disruptive condition of warfare. The casualties of war have frequently involved as many civilians as soldiers. Some historians believe that the Thirty year war in Europe between 1618 and 1648 resulted in the death of 30 percent of the inhabitants of Germany and Bohemia. Despite the combined action of disease, famine and warfare, several major development in the sixteenth, seventeenth and eighteenth centuries stimulated population growth and eased the heavy mortality burdens of man.

The opening of the New World provided a vast new area of space and nature resources of economic exploitation. The industrial revolution greatly increased the productivity and control of infectious diseases. The industrial revolution greatly increased the productivity and mobility of man. These factors combined to accelerate human population growth and the world population curve began to rise upward.

The following table indicates the populations growth rates

Region	Population (billions)			% Annual Growth rate		
	1985	2000	2025	1950-85	1985-2000	2000-2025
WORLD	4.8	6.1	8.2	1.9	1.6	1.2
Africa	0.56	0.87	1.62	2.6	3.1	2.5
Asia	2.85	3.55	4.54	2.1	1.6	1.0
Latin America	0.41	0.55	0.78	2.6	2.8	1.4
North America	0.26	0.30	0.35	1.3	0.8	0.6
Europe	0.49	0.51	0.52	0.7	0.3	0.1
U.S.S.R.	0.28	0.31	0.37	1.3	0.8	0.6
Oceania	0.02	0.03	0.04	1.9	1.4	0.9

(Source: Our common Future, 1987.)

Between the birth of Christ and the seventeenth century, the world's population took approximately 1600 years to double. The next doubling as achieved within 200 years, and the next from 1 to 6 million people within 80 years, from 20 to 4 million people were completed by 1975 a span of 45 years and at the current rates of growth the world population will double again within another 35 years.

Currently the world's population is increasing at approximately 2 percent per year—a net increase of about 1,300,000 per week. Every day of every week, the world's population increases by more the 207,000 people. All this is primarily a product of man's increased life span a reduced mortality through modern medicine and agriculture. Man's fertility patterns were basically established during periods of much higher mortality rates thousands of years ago and they have remained high in the intervening year.

The developing countries least able to afford rapid population growths are experiencing the greatest growth. The population is not evenly distributed around the world. More than half the world's population regions of the total world, i.e. Europe, North America, former USSR and Oceania had only about 22% of the total world population in mid 1988. The remainder was distributed in the developing region of the world. The growth rate is most rapid in Africa. In the coming decade, the population there will rise from 640 million to 950 million and it will be the highest regional growth rates the world has ever known.

According to projection, countries like Ethiopia, Nigeria, Zaora and Kenya will double their populations within the next 20 years. In North Africa the growth rates are slower, but th figures are still rising fast. South Asia and Africa together will account for over half of the increase. Southern Asia's population will rise from 1.2 to 1.6 million. In Eastern Asia the population growth rates have declined rapidly in the past decade, but the number will still increase from 1.3 to 1.5 million people. Several Arab countries have very high population growth rates and the 12 countries in West Asia will increase in their population to 131 million by 2001. The population; of Latin America and the Caribean is projected to increase by 100 million from 440 to 540 million. In Central and South America, most countries have annual growth rates in excess of 2%.

In addition to the large concentration of population in developing countries, the percentage of population under 15 years and over 64 years of age what is called the dependency ratio is also very high. The population under 15 year is about 45% in Africa, 35% in Asia and 39% in latin America, whereas it is only about 23% in North America 22% in Europe and 29% in Oceania. However, the percentage of population over 64 year is higher in the developed regions of the world.

North America has about 11% of its population over 64 years. Europe has about 13% and Oceania about 8%. It is only 3% in Africa and 4% in Asia and latin America. The high dependency ratio in the developing countries means that age group between 15-64 has to support a large population with requirements for housing, clothing, education, health and other social service. The large young population in developing countries also has demographic implications in the near future.

Asia and Africa, and the regions which will account for about 75% of the world's population in the next 35 years or so, and will have the greatest proportion of poor people. In South Asia 51% of the population lives below the poverty line and in sub Saharan Africa 45% Malnutrition, ill-health, illiteracy and the lowest life expectancy, accompanied by inadequate technology and economic poverty, are predominant in South Asia and Africa. It is not very difficult to imagine the miserable plight of the people in their regions in the next century.

The UNEP has warned that continuing high population growth will result in continuing poverty and increasing environment damage and will lead to higher levels of urban growth and international migration. The population report support this that it will be impossible to keep pace with population trends. If the population growth continues as expected, it says within 10 years there will be more than 200 million children who will receive no education.

What does this mean in terms of human welfare and environment quality? It usually means that many countries have excessive demands for products and services which cannot be met. It is exceedingly difficult, if not impossible, for those countries with the highest rates of population growth to meet the requirements of their people for food, housing, jobs and the physical amenities of life, and it is virtually impossible for them to maintain adequate services in education, medicine, public health and social welfare.

Thus we emerge with such problems: (1) 60 percent of the world's people are still inadequately fed, receiving a percapita calorie, intake less than that recommended by the UN (2) 80 percent of the children of India shows evidence of malnutrition; (3) there are only 9 acres of land surface available for every person on the earth, and only 26 percent of this is available for agricultural production; thus only 1.2 acres of potential cropland are currently available for each person and this is decreasing each year (UN States of Hunger Report); (4) many countries, especially in Asia and Africa, still have an illiteracy rate exceeding 70 percent; (5) in many developing nations, there is only one physician for every 5,00 to 10,00 people, a ratio which makes adequate medical care impossible.

This list could be extended considerably to indicate that many parts of the world are currently living under severe population pressure. While the western nation gain strikingly in prosperity, most of the world's people face increasing poverty and population pressure.

When to global picture is considered, the outstanding fact exists that we can not provide a quality life or maintain a quality environment for the majority of the world's people at the present time, let alone provide for more than 1,600,000 new people every week.

Ecologically, this represents a highly unstable condition, it cannot persist indefinitely without major corrective or compensating changes. These grim facts coupled with increasing global communication, rapid social and technological change and rising expectations throughout all nations add fuel to the fire of social and political instability. As noted by Eric Hoffer in The Ordeal of Change:

> "...... a population subject to drastic change is a population of misfits-unbalanced, explosive, and hungry for action. In other words, drastic change under certain conditions, creates a proclivity of fanatical attitudes, united action, and spectacular manifestation of fluting and defiance. It creates on atmosphere of revolution."

The next decade may decide the future of the earth as hesitation for humans. Already our impact has been sufficient to degrade the soil of millions of hectares to thin the ozone layer threatening the rain-forests and to initiate global warming the full consequences of which cannot yet be calculated. At some point in the not too distant future the changes in the environment may cross the threshold of catastrophe.

Again we see the intimate relationships between ecology and human affairs. It does not imply that ecology can provide the answers to all these problems but surely a failure to recognize their ecology background invites certain disaster.

Indian's Population Situation

India's population, as on March 1, 1992 stood at 843.93 million(437.60 males and 406.33 million females). The second largest population country, India is now the home of 16 percent of world's population. With this, every sixth person in the world is now an Indian. The country however accounts for 2.42 percent of the total world area.

Population counts in India in the modern sense started in 1881 since then, a population census has been conducted regularly every ten years. A census of population was conducted during 1871-72 but it was nosynchrononus. Moreover, the 1972 census omitted served territories, the population of which totaled some 33 million in 1881. The population of India as recorded at each decade census form 1901 has grown steadily except for a decreased during the decade 1911-21. In absolute terms the country population has increased by 160.60 million from 680.3 million in 1981 to 843.9 million in 1991. This is ten times the population of Australia and is more than the population of Japan.

An encouraging feature revealed by the census result is the decline in the growth rate of the population in the last decade compared to 1971 to 1981. It marginally decreased from 24.66 percent recorded in 1971-81 to 23.50 per cent during 1981-91. In most state the growth rate declined during the decade. However, Andhra Pradesh, Arunachal Pradesh, Madhya Pradesh, Maharashtra, Nagaland, Jaipur, West Bengal, Daman and Diu, Lakshadweep and Pondicherry which account for one third of the country's population recorded increase in growth rate. Nagaland registered the highest growth rate of the 56.86 percent while Kerala registered the lowest 13.98 percent.

Uttar Pradesh continues to be largest state population wise with 16.44 percent people of the country living there (138,760,417), followed by Bihar with 10.32 percent (86,338,853) and Maharashtra with 9.33 percent (78,706,719). West Bengal occupied fourth place accounting for 80.06 percent (67,982,732) of the country's population followed by Andhra Pradesh with 7.86 percent (66,304,854) Bombay continued to be most populated city of the country with urban population of 12.57 million, Calcutta taking the record place having

10.86 million, Delhi ranks third with a population (urban) of 8.38 million followed by Madras 5.36 million.

One of the important indices of Population concentration is the density of population. It is defined as the number of person per square kilometer. The population density has gone up from 216 in 1891, to 267 person in 1991. In 1901 it was 77 person. Density is the highest in Delhi (6.319) followed by Chandigarh (5,620). Lakshadweep which has the smallest population size is the third most densely populated area with a density of 1,615 person per square kilometer, Pondicherry ranks fourth (1605) followed by Daman and Diu (906) Arunachal is the most sparsely populated area with a density of 1,615 person per sq kilometer. The ten heavily populated districts of the country are Calcutta Madras, Greater Bombay, Hyderabad, Delhi, Chandigarh, Maoe, Haora, Kanpur City and Banglore. All of them have density of above 2,00 person per sq km and 5.0 percent of the country's population lives in these districts. The average density of these 10 districts. The average density of these 10 districts is 5,791.

In any study of the population, analysis of the sex composition plays a vital role. The sex composition of the population is affected by the differentials in the mentality conditions of males and females, sex relative migration and the sex ratio at birth.

Table

Sex Ratio (Females per 1,000 Males) in India, 1901-1991

Census year	*Sex ratio*
1901	972
1911	964
1921	955
1931	950
1941	945

1951	946
1961	941
1971	930
1981	934
1991	929

The sex ratio is defined as the number of females per 1000 males. In India it has generally been adverse to women. The ratio has also declined over the year except in 1981 when it slightly improved to 934 from 930. In 1991, there has been a fall by five points to 929 per 1000 males. But Kerala represented a different spectrum. The State has a higher number of females than males, 1040 females for 1000 males.

In Indian context, a sex ratio of 950 or above can be considered as favorable to females. The states and Union Territories coming under this category besides Kerala are Himachal Pradesh (96), Andhra Pradesh (972) Goa (969) Karnataka (960) Manipur (961) Orissa (972) Tamil Nadu (972) Dadar & Nagar Haveli(953), Daman & Diu (972) and Pondicherry (982) Chandigarh accounted for lowest number of females per 1000 with 793.

Many scholars have inferred from the actual experience of several European countries and the United States of America that populations historically pass through different more or less well-defined stages in terms of their levels of fertility and mortality. As a generalized explanation of process of mortality and fertility decline in those countries, the transition theory refers to a stage of high fertility in those countries, the transition theory refers to a stage of high fertility and high mortality in simple, mostly agarian, societies which finally transforms itself to a condition of low fertility and low mortality after passing through several other stages. Today the transition theory it sought to be equally applicable to the less developed counties which are still in the early stages of demographic change. In effect this theory has come to be increasingly regarded as theory that might anticipate the future demographic trends in countries which are currently having high population growth due to rapidly decreasing death rate and high but more or less stable or only slowly declining birth rate.

There are several expositions of demographic transition theory, but the one dividing the total period into five stage is more commonly used. These stages are as below;

(a) The high stationary stage characterised by high birth rates and high death rates;

(b) The early expanding stage with high birth rates but declining death rates;

(c) The late expanding stage with falling death rates but more rapidly falling birth rates;

(d) The declining stage with low mortality now natality and deaths exceeding births.

The first stage which is characterised by agrarian low income economy is marked by relatively stable birth rate at the maximum fertility level found in societies not using contracepitives, and the death rate is found to fluctuate in response to conditions of prosperity or famines, epidemics and other disasters.

As the economy progresses to become more interdependent specialised and market dominated with the generation of agricultural surplus, the average death rate begins to decline under the impact of better organisation and improved medical knowledge and care; however, the birth rate continues to remain almost stable at the earlier level. This leads to an increasing growth rate of population which is the early expanding stage. As the death rate continues to decline with further improvements in living conditions birth rate also starts declining first slowly and somewhat later at a faster speed making the 'late expanding stage'. In this stage the growth rate starts getting smaller and smaller than what it was in the early expanding stage. In the fourth stage when further reduction in death rate becomes harder to attain, the death rate stabilizes at a low level (between 8-12 per 1000 persons), the birth rate as approaches equality with the death rate and a more gradual growth rate is established when both birth and death rates are low but birth rate fluctuates some what from year to year in response to voluntary decisions of the people.

In the last stage of demographic transition which started to appear in recent years in a few European countries like Germany, Sweden, Switzerland etc., the birth rates has declined even below the replacement level and there may not be enough people in the next generation to replace the present one. For India, we now have a history of around 100 years of changes in birth and death rates.

Table

Birth and Death Rates in India from1891-1901 to 1989

Year	Birth rate	Death rate
1891-1901	48	48
1901-1911	49	49
1911-1921	49	49
1911-1931	47	37
1931-1941	45	33
1941-1951	43	31
1951-1961	44	26
1961-1971	42	20
1971-1981	37	15

1980	33.7	12.6
1981	33.9	12.5
1982	33.8	11.9
1983	33.7	11.9
1984	33.9	12.6
1985	32.9	11.8
1986	32.6	11.1
1987	32.2	10.9
1988	31.5	11.0
1989	30.5	10.2

The first stage of demographic transition in India continued till about 1920 when both death rates and birth rates were high. This stage was also marked by purely agrarian economy with hardly any significant industrialisation. High mortality levels were hardly and for the variable growth rates of the population during the 30 or more years before 1921.

A severe famine affected large areas of the country in 1896 and 1897. In the Bombay Presidency (excluding the princely states) especially, the effects of the famine were aggravated by a severe plague. As a result of the two scourges, the 1901 census recorded a population loss of about 2 percent from the 1891 level of 18.8 million. Like the 1891-1901 decade the next decade also witnessed several local famines and a severe one in 1907 in most parts of Uttar Pradesh. Plague was in evidence in the Bengal and Bombay Presidencies, and both plague and malaria were widespread in the Punjab and Uttar Pradesh where population growth was negligible. Yet, because the country as a whole suffered from famines less widely and for shorter durations, there was an appreciable rise in the population growth rate compared to the previous decade.

During the 1911-20 decade India suffered from an influenza epidemic that caused an estimated 7 percent of the total population to die and led to an absolute decline in the population of the Indian Republic (according to the present boundaries) by 0.77 million at the 1921 census. The second phase of demographic transition in the country started with the beginning of the 1920s and has spread well up to 1971. This is also the period when industrialization began in the country and took some roots. During this period the major causes of high mortality-famines and epidemics-have been gradually brought under control and between 1921 and 1951, the country witnessed a gradual rise in population growth rates.

The decline in death rate became sharper during the 1950s without any appreciable decline in birth rate resulting in a population growth rate of 1.96 percent per annum during 1951-61 in comparison to 1.25 percent during 1941-51 decade. During1961-71 the birth rate still remained quite high and due to further decline in death rate the growth rate increased to 2.05 percent.

India seems to have entered the third stage of demographic transition since 1971 when birth rates has also started declining. During the birth rate leading to a plateau in the population growth rate during the 1960s and the 1970s. In the 1980s, probably the decline in birth rates was slightly greater than that in mortality; consequently the growth rate during this period is slightly lower than in the previous decade. It is now hoped that there would be faster declines in birth rate than in death rate in the next couple of decades. It is also hoped that India will be at the point by about 2020 if not earlier.

Only one in ten people lived in cities when the century began more than half will by the century's end. Today the vast bulk of urban population growth occurs in developing countries. The populations of Third World cities, now doubling every 10-15 years, overwhelm government's attempts to provide clean water, sewage, adequate transport and other basic services. Cities in Industrial countries are gaining people much more slowly, yet even as their growth rates slow or core population decline, these urban areas continue to spread out ward.

Industrial and developing countries alike, chaotic, uncontrolled urban growth-whether measured is risking numbers or in the amount of space humans spread out upon—draws on ever more land, water and energy from surrounding regions to meet people's needs. The way cities physically evolve—and the way their development is planned has profound impact on human either recognize the limits of the natural environment or it can destroy the resources on which current and future societies depend, it can meet people's needs equitably or it can enrich some while impoverishing or endangering others.

The world needs an urban planning ethic that is sensitive to these environmental and human dimensions. According to 1981 census, India's urban population was 160 million ranking fourth among the countries in the world in terms of absolute size of the urban population, the first three being China, U.S.S.R. and U.S.A. Though the level of urbanization measured as the percentage of urban to total population has increased from 11 percent in 1901 to only 23 percent in 1981 the number of people living in urban areas and multiplied six folds during the period from 26 million, constituting more than one third of the country's population of 926 million at that time.

The growing chaos in urban areas are due to rapid immigration rather than natural increase. A total of 15.7 million persons have migrated from rural to urban areas in the country during the decade 1971-81. The immigrants in the reverse direction i.e. urban to rural numbered. 6.4 million giving rise to a net addition of 9.3 million people in urban areas.

There are 218 class I cities in India with more than a lakh population each and these account for about 60 per cent of urban population. Among these, there are 12 metropolitan cities viz., Calcutta (9.16m), Greater Bombay (8.22m), Delhi 5.7m), Madras (4.27), Bangalore (2.9m), Kanpur (1.68m), Pune (1.68m), Nagpur (1.29m), Lucknow (1.0m) and Jaipur (1.0m). With the data on class I cities and the emergence of new towns on the eve of the 1991 census, we can examine the likely level of urbanization in India. It is well known that the share of Class I urban population has constantly agglomerations and cities in the total urban population has constantly increased over time.

If one looks at the total urban population has constantly increased over time. If one looks at the trend in India (excluding Assam) from 1951 to 1981, this share increased from 44.7 per cent to 60.8 per cent, the increase being 6.8 per cent during the 1970s. The increase in the share was higher during the 1950s since a very large a number of small towns (857 in all) were declassified on the eve of the 1961 census due to introduction of a restrictive definition of urban areas.

In contrast, there was a net addition of 895 towns (or 782 urban agglomerations and towns) between 1971 and 1981 which involved reclassification of almost 250 towns declassified on the eve of the 1961 census. Addition of a large number of small and medium towns during the 1980s brought down the pace of increase in the share of population of Class I cities in 1981. As the number of new towns indentified on the eve of the 1991 census is quite large, one may expect a substantial increase in the population of medium and small towns in 1991. If, however, a large number of them appear as part of urban agglomerations of Class I cities, their population will be counted as population of Class I cities. An examination of the list of new towns and of cities and urban agglomerations indicate that a large number of new urban settlements have appeared as part of urban agglomerations.

Accordingly, we may expect the share of the population at 65 per cent. This is in view of the fact that 80 Class I cities have been added during the 1980s. We may also not forget that the decadal growth rate of Class I cities slightly declined over the decade compared to the previous decade. Assuming that the proportion of population of Class I cities in the total urban population of 1991 is 65 per cent, we get the total urban population as 213 million excluding that the Assam and Jammu and Kashmir. If we add the 1991 population of Class I cities of Assam, the estimate of total urban population would remain at the 1981 level (that is, 60.5 per cent), the 1991 urban population (excluding Jammu & Kashmir) works out to 230 million, almost the same as projected by the Expert Committee on Population projections in 1984. Thus, the urbanization rate in 1991 varies between 25.6 per cent to 27.5 per cent.

Keeping in view that the number of metropolises has increased from 12 in 1981 to 23 in 1991, and that the number of Class I cities has increased from 216 to 296 during the same period, one may expect a higher share of Class I cities in the total urban population than 65 per cent. If this is found to hold good when the data on rural and urban population are released, the total urban population would be less than 214 million.

Population in the twentieth century has exploded particularly in the developing countries. Infact, before 1800, no country was predominantly urban. But by the year 2000, over 50 per cent of the world's population will probably live in urban places and there will probably be more than 250 cities of over one million inhabitants. There has been a world wide trend towards urbanization, the process characterized by the movement of people from rural to urban settlements, from small towns to large cities, and from large cities to the suburbs. This immigration is caused by and in turn leads to profound societies change. Rural areas are depopulated, while urban areas towns, cities and urban zones—become more densely populated. Since most of the cities are growing in an unplanned way there arises a number of problems such as housing, traffic, shortage of open space & pollution.

The Indian cities are growing both vertically (multi-storey building culture) and horizontally (invasion of the peripheries). In spite of cities expansion there is growing shortage of house (3.8m), growth of slums and squatter settlements, creating interpersonal and intercity imbalances, breaking down services and acceutuation of the problem of urban renewal. Slums are now becoming permanent features and are proliferating and sheltering larger population every year. They are having small tenements, narrow lanes, lack of sanitation, water supply, electricity, unhygienic environment, etc.

Due to realization of the futility of spending money on their clearance, the Government had decided to accord them recognition and gave their residents whatever amenities the funds allowed. Bombay is estimated to have the largest slum population of 3.3 m followed by Calcutta 3.2m, Delhi 2.6m, and Madras 1.8m. With an unabated inflow of job seekers, these cities are expected to see continuing rise in their slum population. By 1990 when Calcutta will have a population of 12.5m and its slum will be harboring 4.4m Madras population will be 11.8m, 9.7m and 6.0m and of them 4.1, 3.1 and 2.1m slum dwellers. Amongst States, Uttar Pradesh will have the largest slum population of 6.5m followed by Maharashtra 6.2m, West Bengal 4.9m, Tamil Nadu 4.2m, by the end of the decade. The Government's initial strategy was to clear the slums and rehabilitate their dwellers. This had to be given up by the end of the sixth plan as there was constant shortage of funds and developed land.

The emphasis was shifted to an on going central scheme known as "Environmental Improvement of Urban Slums," envisaging a package of services for the immediate removal of unhygienic environment of the slums. The services included drinking water, community bathrooms, street-lighting, sewerage, storm water drains and street paving.

The traffic problem in urban areas is one of the greatest problems of the day. Most of the cities and towns are growing in unplanned way. Due to lack of proper enforcement agencies encroachment in all parts of the cities is going rapidly and due to political interference, authorities become in competement. Most of the roads are narrow, haphazard and ill-maintained. Traffic problem is getting bad to worse because of high density of population desire to have personal means of transport and to live near the city, mixed traffic, lack of civic sense, ill-maintained road, inadequate expansion in the existing road network poor regulation of traffic, etc. Delhi had 12,000 vehicles in 1951 and the number had shot upto 9 lakh in 1986 and this is expected to reach 40 lakh by 2001.

Apart from the fast moving vehicular traffic, the problem on the city roads is being further aggravated because of 50,000 or more slow moving vehicles such as tongas, rickshaws and bullock carts and 20 lakh cycles. There is also a growing trend of road accidents. There were 6,037 reported road accidents of which 1,162 persons lost their lives in 1986. According to National Transportation Planning and Research Centre, Trivandrum in every four minutes there is one road accident in India, killing or injuring one person on an average.

Due to increasing commercial value of the land in cities, open spaces are diminishing, even though green spaces are the lungs of the cities, the guarantors of a healthy and aesthetic environment. In Japan, urban parks are classified under two main categories; parks with structures or city parks and zoned green spaces, which are further subdivided into green space conservation zones, agricultural green space and special historical land space zones/scenic beauty conservation zones.

City park land includes playgrounds, open spaces, planting strips along side road and other lands awaiting full development. It is commonly said that these parks become the pursuit of undesirable activities including sex, crime, gambling, etc. In spite of these negative aspects, it provides thousand times much needed benefits for the city dwellers. There is a concept to have at least 20 per cent area of the cities under open space in the form of green belts, parks, playgrounds, etc. But it has become an utopian concept. In 1973, master plan of Allahabad was approved and much of the areas were put under green belts. But due to illegal spraw of colonies most of that very areas were converted into concrete belts.

The green belts concepts are also not getting proper place in the city environment improvement scheme because of lack of financial support, public interest and professionally qualified staff. Besides this, there is not definite framework for parks and recreational areas in relation to national goals. City parks and green belts are developed mostly by Municipalities or Corporations,

Social Forestry and Horticulture departments. Private organizations, voluntary organizations, industries should also come forward to set city parks and green belts. Vacant lands acquired by the Corporation are to be sued not only for commercial or residential settlement but also for establishing big parks. Local communities must come forward to participate in the management of parks and can also raise funds for it.

Country parks can be established in rural areas to provide a better place for recreation and enjoyment of rural people. In addition to parks and gardens, landscaping measures should be applied on flyovers, overhead bridges, wall and other structures, such as the growing of creepers, climbers, scrubs and transplanted trees. There is a need for making legislation for City and Country Parks Act for the forceful implementation of the concept of green belt.

Virtually all cities share some land related concerns, such as congestion and pollution from motor vehicles, lack of affordable housing, and the cancerous growth of blighted district. Of course, prescriptions for better urban planning are not the same for all parts of the world. Many land use issues in industrial countries seem mainly concerned with the quality of urban life, and those in Third World cities are often questions of life and death. But all cities, whether surrounded by affluent suburbs or makeshift shanty towns, now need to plan land use far more carefully than in the past before the developing world's urban crises than into catastrophes, and the industrial world' urban crises turn into catastrophes, and the industrial world's problems become issues of survival.

The extent to which cities control the use of land varies widely. Japan and Western Europe have the world's most comprehensive urban land use controls; their cities are generally compact, laid out for efficiency in transport, water and energy use, and building materials. Japan protects farmland with strict zoning and tax policies. Its suburbs are of moderate, not low, density. Much of Japan's urban activity is concentrated in Tokyo and along the dense Tokadio corridor between that city and Osaka. This compact, linear form has allowed great efficiency in transport, including development of the world's first high-speed train, which has been running since 1964. However, the country's 33 million automobile—nearly twice as many as in all other countries combined—still plague Japanese cities with extreme traffic congestion, especially in the Tokyo area.

Western Europe has a long tradition of activity controlling land use so that the small amount of available space serves the public's interest more than that the private developers. Paris, a city for nearly 2,000 years, has planned and regulated land use since the Middle Ages. Many of England's urban areas, still fulfilling a far sighted decree by Queen Elizabeth I in 1580, are ringed by green belts intended to protect farmland and prevent sprawl. But in post war decades the automobile has had a profound effect on Europe's compact character, fuelling suburbanization. In the fifties and sixties, some older European cities embraced the U.S. model, demolishing corner shops and cafes to make way for urban freeways.

Later, planners tried to contain suburbanization by diverting growth into smaller satellite towns. Since the mid-seventies, they have turned their attention to revitalizing shrinking inner cities. Control of land use in Eastern Europe and the Soviet Union has not been as effective as might be assumed for centrally planned economies. Urban planning therefore is not well coordinated with economic decisions, and sometimes reflects rivalries among government ministries. Extreme water air, and soil contamination from poor placement and regulation of polluting industries is a critical concern. Although urban planners in these countries historically have supported public transport and promoted urban densities high enough to make it viable, in recent years and use has become less efficient. Nearly all new housing has sprung up at the edges of cities, with few accompanying shops, jobs, or schools, resulting in excessively long commutes.

On going political reform and the region's cities will become more automobile-oriented, perpetuating the environmentally destructive patterns of western industrial cities. Among industrial regions, North America and Australia have the weakest planning traditions. Governments on these continents have done relatively little to guide development beyond separating industrial areas from those zoned for commerce and housing. Many have enforced low-density zoning in an effort to curb urban expansion—a move that, ironically, achieves just the opposite.

Zoning codes that restrict residential density, usually by requiring each house to occupy its own large lot, have forced development to consume even greater tracts of open space. And to accommodate new, outlying communities, roads have been extended further outward, attracting even more spread out growth. Even older U.S. cities are becoming more dispersed now than in the pre-automobile area. Growth in the New York metropolitan region has turned to sprawl; while population has grown only 5 per cent in the past 25 years, the developed areas has increased by 61 per cent—consuming nearly quarter of the region's open space, forest, and farm land.

Many people move out to the suburbs and exurbs seeking open space and bounds with nature that come only in a rural setting. Yet most of these residents continue to maintain urban lifestyle—communiting to jobs in the city and demanding urban amenities from a suburban shopping mall near their home. The result is neither urban nor rural living but a destructive compromise that the environment sustain. Developing countries have the loosest controls over how cities develop. Local governments in Asia, Africa, and Latin America often have neither the authority to guide land use nor the funds to provide basic services. Municipalities are fiscally dependent on central government, which are themselves struggling under crushing debt burdens.

With few exceptions, urban planning is a relatively recent phenomenon in developing countries, even in the capitals. In India it reportedly began with Delhi's Master Plan in 1962. In Nigeria, the first systematic effort to plan cities did not begin until 1971; the capital, Lagos, still has no designated metropolitan government and no formally defined urban area. The Lagos State Ministries that have their own priorities. The Third World is burdened by several enormous, rapidly growing cities—including-Sao Paulo, Shanghai, and Mexico City—whose sheer size and instability create problems on an entirely different scale. As early as 1964, researcher Ronald Wraith used a single word to denote these giant cities racked with pollution and rimmed by shanty-towns: "Megalopolis-the city running riot with no one able to control it." Since then, megacities have become increasingly characteristic of the developing world. In 1950, only three of the world's 10 largest cities were in the Third World by 1980, seven of them were there.

Although skilled land use planning is badly needed in these megacities, its effectiveness is limited in the absence of other kinds of change. One reason is that most of the physical growth takes place in illegal unplanned squatter settlements, rendering useless even existing mechanisms for guiding land use. Called fevelas in Brazil bidonvilles in French-speaking West Africa, Spanish in the Middle East, and kampungs in Indonesia, these illegal communities hold 30-60 per cent of the population of many Third World cities. The unhealthy conditions in these settlements can only be addressed fully through extensive economic and social reforms that attack the root causes of poverty not just in cities but also in the rural areas that urban migrants abandon in search of economic opportunity.

The effectiveness of land use planning is also limited in the absence of family planning programs. (Only in Africa does migration account for the increase in much of Asia and Latin America results from the fertility of people already in cities). Finally, planners in the Third World's giant cities faces colossal environmental problems, from deadly air pollution in Mexico City, to indiscriminate dumping of toxic wastes in Alexandria, to the actual sinking of cities such as Bangkok, Jakarta, and Shanghai due to over-drawing of ground water.

Crowding is a conspicuous ecologic force impinging upon modern man. It results from more than a simple increase in population density. It is also a product of activity patterns psychological needs, and social contact. Crowding implies a force, a pressure and psychological reaction. No clear definition of crowding is possible at the present time; it remains an operational concept with many subjective qualities. Crowded population are, however, easier to recognize than to define, and they are studied more carefully, better definitions may become possible. For the time being, we might define crowded populations as those in which there seem to be excessive number of individuals per unit space in relation of the activity of the industrials and the quality of the environment.

With an operational concept of this type, ecological studies have shown that some rodent populations may be "crowded" in relation to their social behavior and habitual requirement at a density of 5 individuals per acre, where as other at the same species may not be "crowded" until they have a density of several hundred per acre. Species differ in this regard, and so also individuals and social groups within the same species.

A western cattle rancher may feel "crowded" at a density of 10 homesteads per square mile whereas a modern suburbanite may feel relatively "uncrowded" at a density of several hundred homes per square mile. The western rancher ranges several square miles per day over what he considers his personal property and he does not welcome potential competitors in his territory. Conversely, suburbanite restricts his daily movements, apart from going to work and to shop, to a lawn around his home, and he does not view neighbors as a competitive threat. Hence, in both animal ecology and human sociology, it is important to distinguish between population density per se, and crowding, since the latter depends so much on behavioral factors.

The spatial requirements of man are obviously very variable, and result from both individual difference and social conditioning. It is impossible to state what the optimal spatial and social needs of man are as those of the Japanese city dweller ar much different than those of the Kansas farmer. Nevertheless, it is safe to assert that man is now subjecting himself to increased crowding through the combined forces of population growth, urbanization, transportation and mass communication.

Man's perception of crowding has skyrocketed upwards in recent years. The growth of urban traffic and the increased number of people in public buildings, schools, hospitals, stores, exhibitions, recreational areas, etc., have all contributed to the feeling of population pressure. Through the mass media or audios, television and newspapers, individuals are vastly more aware of worldwide events. Social pressures at any point of the globe are quickly felt by all nations. Local crime, violence, and warfare now become worldwide worries in a matter of hours. The people of one continent feel threatened by those of another continent thousands of miles away.

The inevitable result has been a great magnification of social anxiety and population pressure. The world has thus become more limited in its psychological space, and man more crowded in a social and behavioural sense. How does one evaluate crowding as an ecologic force? It definitely requires knowledge of at least three major parameters: (1) population density, (2) environmental structure, and (3) social behaviour in relation to each of the possible in both animals and man, but in both, especially man, it is often possible to isolate crowding as the key variable.

Many other factor are interwoven. Crowded human populations are often so characterized by poverty, malnutrition, lack of educational and recreational opportunities, varies environmental hazards and unstable social patterns that it is virtually impossible to separate crowding itself as the one key variable. The essential question becomes, how does one measure and analyze crowding per se apart from all it related phenomena? It may also be important to distinguish between crowding of various origins. Since man often seeks social contact and stimulation, does not crowding contain constructive and creative social elements, or is it inevitably a factor in increased social problems and environmental deterioration?

From man himself the answer to these questions are confusing and uncertain. The behavioral sciences of man have just begun to investigate human spatial behavior. More direct clues have come from animal studies where experimental work has been possible. The simpler social systems of many animals have permitted more direct observation of behavior, more ready identification of variable, and more penetrating analyses of causes and effect relationships. In animal studies it has been possible to distinguish between extrinsic crowding; that crowding imposed on the animals as an external force, as in captivity, compared to intrinsic crowding in which dense aggregations result from the social behavior of the animals.

A similar distinction is sometimes possible in human populations, but frequently it is difficult to discern if people are crowded by choice or the external pressures of economics and society. It was demonstrated more than 30 years ago by Professor W.C. Alee and his students of the University of Chicago that most species of animal have optimal levels of crowding, above which and below which deleterious effects occur. These studies showed that certain levels of crowding are necessary to maintain normal biologic function and social activity. Goldfish, for example, grow most rapidly and live longest in social groups of certain densities compared to isolated fish to fish in smaller groups.

Fresh water shrimp survive longest and tolerate harmful environmental conditions most successfully at certain optimal stage of crowding. Similarly bobwhite quail survive harsh winters best and reproduce most successfully in the spring in conveys of certain optimal size. Beyond optimal levels of crowding, however, benefits are quickly lost and various forms of pathology, occur. Growth may be stunted, life span shortened and various diseases or abnormal behaviors appear.

Most studies on crowded animal populations in the last 20 years have concentrated primarily on the harmful effects of overcrowding. Such effects usually involve both physiologic and behavioral abnormalities. The physiologic effect of overcrowding centre around endocrine imbalances of the adrenal pituitary system. These imbalances can lead to a great variety of disease condition including gastric ulcers, hypertension, nephrosclerosis, arteriosclerosis, and increased susceptibility of infectious diseases.

Prolonged exposure to crowding and its associated stresses have produced additions disease (hypoadrenalism or adrenal exhaustion) myasthenia (muscular weakness), and a variety of serious metabolic disorders leading to shock, coma, and even death. It should be pointed out that these stress effects are quite variable, however, and do not consistently reflected the same patterns in different individuals and various social groups. Behavioral pathologies also appear in experimental animals as a result of crowding and these are sometimes manifested more quickly than are the physiologic changes.

Overcrowded animals often show increased aggression and violence, abnormal sexual behavior, disruption of normal nest building and maintenance, breakup of stable social groups, disappearance of alteration of normal social roles, parental desertion of young and frequently cannibalism of young. The leading question, of course, is whether or not similar processes occur in man as a result of overcrowding. There are so many variables in man which affect health and behavior, that this becomes a very difficult question. There is no doubt that some striking correlations of disease and behavior are associated with urban concentrations. Many human problems seem to be so accentuated in crowded populations that they virtually constitute an "inner city syndrome" disease, and emphysema; maternal problems such as high prenatal loss, parturitional difficulties and high infant mortlity; behaviour problems such as alcoholism, drug abuse, mental illness, and criminal assault; and social problems such as high divorce rates and social instability. More recently it has been found that high prevalence of mental illness in the lower socioeconomic classes and dense urban ghettos of several other cities. The three volume report on Manhattan entitled, "Mental Health in Metropolis," also found unusually high prevalence of mental illness in crowded inner city inhabitants.

The problem of urban man seem so strikingly similar to those effects produced in overcrowded animal populations, it is tempting to conclude that similar processes are occurring. But, as stated previously, many factors are interacting with crowding in urban manpower, malnutrition, lack of educational and recreational opportunities, various environmental hazards, and unstable social patterns—that it is virtually impossible to separate cause and effect relationships. There are enough solid exceptions to the principle of "high density produces high pathology" that one must be cautious in carrying this dogma too far. For example, the young nation of Singapore is obviously crowded—over 2,000,000 people on an island of 224 square miles, making it one of the most densely populated nations of the world—yet it enjoys remarkable good health.

The death rate is one of the lowest in the world (5.5 per 1000) and the prenatal mortality is lower than that of Britain and the United States. Major factors in this good record many be its youthful population (40 per cent of the Singapore population is under 15 years of age), and the well planned interspersion of parks and greenbelts with residential and commercial areas. One does not feel a "crowded" in generalizing as in Hong Kong, Tokyo, New York and Calcutta. Such variable demonstrate the difficulty of generalizing broadly about crowding or population density as a pathologic influence. Thus, the precise role of crowding in the "inner city syndrome" must remain a cloudily issue until more and better research studies are available. An encouraging development of recent years is the awareness on the part of some architects, city planner anthropologist and others of the relationship between our environment and mental health.

In Architectural Environment and our Mental health, Moiler explores the impact of physical space and structure on the physiological well-being of man. He noted that more buildings will be built in the next ten years than have accumulated since the beginning of civilization." And he wisely emphasized that, "We should recognize that if we continue to permit the structuring of a physical environment which is essentially hostile, to which man must somehow try to adopt, the cost to individuals and to society will be far too grat—in terms of mental illness, delinquency, poor motivation and fulfillment of our capabilities for creative work and community usefulness." Anthropologists have also been interested in the spatial behaviour of man, and the book by Professor E.T. Hall entitled "The Hiddeen Dimension," was the first one to highlight some of the behavior patterns of man relating to social spacing. Hall showed the existence of "personal space" individuals and pointed out cultural differences in the extent and importance of personal space.

The process of social and economic development involves a variety of interrelated factors which lead to change. For example, a marked reduction in birth rate in a country or region may start chain effects, on a number of aspects of socio-economic life. It affects the age structure of the population and reduces the period or young dependency. It may increase the education of women and induce a higher proportion of women to participate in the labour force changing the traditional role patterns of women as housewives and mothers

The total development of a country or nation is a result of changes in several interrelated factors like rise in the income level, greater availability of food, enlightened attitude towards women's education etc. It is seen that one sided emphasis on economic development or the sectoral approach to development usually leads to undesirable consequences. For example in Kenya though the rate of economic development has been quite high by any standard (an annual increase of 6.4% in the gross domestic product during 1960-70), the economy has made the rich richer and poor poorer. Further, a rapid economic development on the one hand and slow social development on the other has created a piquant situation yielded in recent years in the petroleum producing countries of the Middle East.

In Iran, it is noted that the rate of increase in GNP (gross national product) during 1971-72 was 14.2% where as it was 30.3% in 1972-73. Also Iran's oil revenue in 1973 was just 5 billion dollars which phenomenally rose up to 25 billion dollars in 1975. High economic growth has been achieved in Iran comparatively in a short period of time. But in spite of its best efforts, Iran has not been able to achieve the same spectacular progress in the field of social development. It may therefore be concluded that development is a very complex process and there is no simple input-output law of relationship governing the total development of the country.

The relationship between population growth and food supply in South Asia in general and in India in particular has often been regarded from a neo-malthusian perspective. Needless to mention that after independence the food production has increased more than threefold. But even this quantum is inadequate for the present population. In fact, India possesses about 2.4% of the total world land area with which she has to feed about 16% of world population. In spite of the increase in food production a large section of the population, remains undernourished mainly because of the high increase in the mouths to demand food. It has been postulated that protein deficiency is indirectly caused by a low calorie intake. A number of studies have been conducted on the relationship of food production and per capita availability.

In many countries, including India the yield per hectare hag almost doubled within the last two decades as a consequence of Green Revolution, use of fertilizers, irrigation, mechanised cultivation etc. The effect of Green Revolution, is mostly visible in the area of wheat harvesting. The produce of wheat in 1960-61 was 11.00 millions tonnes which rose in 1975-76 to 28.33 million tonnes. However, the Green Revolution remains confined to areas having adequate irrigation facilities such as Punjab, Haryana and some parts of Uttar Pradesh. About 70% of India's population is involved in agriculture and about 40% of Indian economy depends on agricultural produce. It will not be an exaggeration to say that the food situation in India which depends on the vagaries of monsoon has usually been critical. Per capita cultivated land has been shrinking in size despite the marginal increase in the acreage of cultivable land through land reclamation. Between 1921 to 1981 per capita cultivated land dropped from 1.11 acres to 0.62 acres.

Even concerted efforts to increase the productivity of land by means of science and technology and crop rotation, did not succeed in raising the level of per capita domestic availability of foodgrains in any significant way in the last 30 years the increase has not been more than 8%, mainly because the increase in food production is largely offset by population increase.

Increase in per capita food availability is one of the indicators of the improvement in the standard of living of the people in developing countries, where per capita caloric intake comes largely from cereals and pulses and such calorie intake is unenviably low in comparison to developed countries.

India comes in the second phase of demographic transition. On comparing the demographic situation of India with those of developed nation it is clear that India's birth rate should decline without any delay. Looking upon the future prospects for population and food supply in South Asia, it is seen through the 'medium variant' United Nations Demographic projection that South Asia's Population by 2020 would be 1881 million. Of the total increment of 750 million an increase of 482 millions is projected to occur in India. On the other hand, if per capita cereal consumption stays constant at 1990 levels, then total cereal demand for South Asia will also rise by 66% to reach 427,180,000 metric tons in 2020 approximately double of current levels of consumption.

Further it is projected that present increase of 2.3 in food production in India will remain 1.8 by 2001 while population growth will be 2% indicating high load of population pressure on the available food production.

The literacy rate is a key to every type of development and therefore each of India's development plans accords a high priority to education. Since population growth and education are closely interrelated. Efforts to enhance the literacy rate in the developing countries is very much affected by the increasing growth of population. For example, even if the fertility is substantially reduced, the number of school-age children will increase significantly. It is expected that in developing countries the increase will be much greater i.e. 80% by the year 2000, assuming a decline in fertility. If the number of children increases more rapidly than the population, the need for educating will claim a large chunk of the national resources. In fact such countries where the population of school-going children is large are educationally backward. In developing countries the age structure of the population becomes a barrier to educational development. As a consequence per capita allocation for education becomes smaller.

In India in contrast to the developed world, age-wise population distribution has a pyramidal pattern where children constitute 39.7%, youth 32.6%, middle age persons 21.5% and old age people 2.67%. So, India has not been able to impart education to all children. Among various contributory causes for this state of affairs the most important one has been the yearly increase in the number of new entrants to schools already saddled with the back log of yesteryears. This situation creates problems for educational planners. For the reason that school buildings classes are allowed to get joint sized. Even the same building is used for various shifts.

The net result is that school buildings are unexpandable, although small and inadequate student-teacher ratio goes up affecting adversely the quality of education. In India, despite the substantial increase in educational facilities, the target fixed for elementary education has not been reached anywhere near it. It may be observed that from 1950-51 to 1968-69, there has been progressive improvement in the enrolment of the children in the age-group 6-14 in classes I to VIII.

The Govt. of India's expenditure an education is only 2.5% (1980-90) less than every many south Asian countries (Pakistan 2%, Nepal 10.9%, Bangladesh 11.2% and Sri Lanka 9.9%). To educate the growing population India needs about 2,00,000 more primary schools by 2001 and more and more modern technology, staff, buildings and related materials for higher education. Again it will be a problem under the available resources. The enrolment of girls in schools, however, has not been very satisfactory. According to 1991 census of India only about half (51%) of the population of India is literate consisting 63.8% males and 39.4% females.

Rapid population growth also raises the number of persons in working age-group. This leads to migration of the people from one region to another. Although, the growth rate of gross national product has registered constant upswing over the past few decades, the developing nation's economies have not been able to absorb fully the growth of their labour force. Poverty and unemployment are the two biggest problems for India which have their implications in the economic, social and political realms of life.

Any discussion on the development would, therefore, remain incomplete without touching the problem of unemployment. In India, the children instead of going to school, help their parents on farms and also share household work. In some cases they work even as labours to supplement income of the family. According to 1971 census of India about 72% of the population was found engaged in agriculture. That was much more than what obtained in countries where this percentage Is less than 20. In most developed countries children upto the age of 15 years are not permitted to work but in India quite a significant percentage of boys (9.4%) and girls (6.6%) are found engaged in some kind of work or the other.

In all countries there is an increasing trend towards unemployment in spite of increasing opportunities for employment. In industries and other development projects. The percentage of the unemployment in India is not available but from the total number of people unemployed in 1962 and 1972 it is evident that unemployment has increased about three times and among educated people it is rather alarming. Many talented youngmen in the field of engineering and technology, are either unemployed or underemployed, which is a great national loss.

The availability of Health Services is another indicator of the quality of life of a given society. The availability of health facilities is, directly linked to the number of people in a country as much the status of the economic development. Health as defined by the World Health Organisation is a state of complete physical, mental and social well-being and not merely the absence of disease or infirmity. According to the recommendation of Health Survey and Planning Committee (Mudalier Committee), there should be one doctor for every group of 3500 persons, and one hospital with adequate beds for a population of 1000. In developing countries the number of users of a hospital bed ranges between 330 and 3700. In India the number of possible users for every hospital bed was 1671 in 1965.

The population per physician is also low in developed countries. It ranges between 400 and 900. In developing countries a physician is required to attend to any number of patients between 1500 and 49000. India had 4610 people expected to be looked after by one physician in 1969. Central Govt. of India's expenditure on health during 1980-90 was 1.6% less than Bangladesh (4.8%), Nepal (4.8%) and Sri Lanka (5.4%). In India, there is only 2.1 hospitals per 1,000 sq. km. However, it becomes difficult to have "Health for all by 2000 AD". In any developed country approximately 100 doctors, 250 nurses and 750 hospital beds are available for serving 100,000 population. To achieve this goal, India needs about 7,00,000 doctors, 23,00,000 nurses and 70,000,000 hospital beds. This seems difficult to achieve on the basis of available infrastructures.

The Five year plans of India have made increasing allocation for health services. In the First Five Year Plan the outlay was Rs. 140 crores, it was Rs. 225 crores in the Second Plan. It was in creased to Rs. 2334 crores in the Fifth Plan. By the end of the First Plan, there were 65,000 doctors in the country. This figure went up to 1,38,000 by the end of the Fourth Plan. The number of hospital beds also went up from 1,25,000 in 1951-56 to 2,81,600. There were 42 medical colleges in the country by the end of the First Plan which went up to 1000 by the end of the Fourth Plan. At the present rate of population growth, economic conditions and the facilities for training of physicians in India, it seems very likely that the health services will deteriorate instead of improving. At present about 2000 medical undergraduates are churned out by 99 medical colleges in India. Out of this some continue there studies at the graduate level and some emigrate to other countries. Hardly about 1500 undergraduates remain available every year for service in hospitals in India, which meet only one-half of the demand of the existing population to maintain physician-patient ratio. The chances of improving the situation are thus very bleak unless the growth of population is drastically reduced immediately.

Population growth of individuals depends on physical factors as food, water, space and energy for their survival. The scarcity of any of these factors can become a 'limiting factor' for the future growth of a population. The earth resources are getting scarce because of increase in population as well as increase in consumption of different resources by more and more people in order to improve their standard of living. In an effort to save humanity from misery and improve the quality of life man has been polluting the environment. The relationship of environmental pollution can be directly as well as indirectly linked to increasing population. Rapid consumption of natural resources and the waste disposal by increasing number of people put a greater stress on the environment. The rate of consumption of developed countries is much higher than that of developing countries. Because of this higher level of consumption in the industrialized nations even relatively modest increase in the population can result in far greater stress on the environment. The pollution problem is felt more in the industrially developed countries than in the developing countries.

Three major factors determine future environmental trends. These are (i) economic growth (ii) knowledge and (iii) social technology. These may not be achieved under a constant pressure of high population growth. The population of India reached the 685 million mark in 19811 and is currently growing at the rate of nearly 1.3 million persons per month. About 23 million babies are born each year and about 8 million persons die, bringing about a net annual increase in population of 15 million, numerically equivalent to the population of Australia.

India is the second most populous country in the world after China. The population of China, enumerated on July 1, 1982 was 1008 million. Thus, India and China together had a population of around 1.72 billion in mid-1983 which is more than one-third of the world population and roughly half the population of the world's less developed regions.

Among the States in India, the largest number of persons were enumerated in Uttar Pradesh—around 16 per cent—followed by Bihar with around 10 per cent. Nagaland is inhabited by the least number of persons and accounts for only 0. 11 per cent of the total population of India. Among the Union Territories, Delhi contributed the highest proportion and the lowest number was enumerated in Lakshadweep. Only nine

States, namely, Andhra Pradesh, Bihar, Karnataka, Madhya Pradesh, Maharashtra, Rajasthan, Tamil Nadu, Uttar Pradesh and West Bengal, contributed more than 5 per cent of the total population of India. The States where less than I per cent of the total population was enumerated are Himachal Pradesh, Jammu & Kashmir, Manipur, Meghalaya. Nagaland, Sikkim and Tripura. These States stretch along the foot of the Himalayas.

India accounts for nearly 15 per cent of the world population but occupies only 2.5 per cent of the total land area. Thus, compared to the average density of the world, India is densely populated. The average density of population in India in 1981 was 224 persons km^2, which is approximately double of what it was in 1951. Among the States, Kerala and West Bengal are the most densely populated States in the country where on an average more than 600 persons were enumerated in every square kilometre area.

Nagaland and Sikkim are the most thinly populated states with a density of less than 50 persons/km^2. Apart from Kerala and West Bengal, Assam, Bihar, Haryana, Punjab, Tamil Nadu and Uttar Pradesh are densely populated States where the density is found to be higher than at the national level. Among the Union Territories, Chandigarh and Delhi are the most densely populated areas where on an average more that 4.1 thousand persons were enumerated/km^2 area. These two Union Territories are highly urbanised places and in fact have the highest density in the country. On the other hand, Lakshadweep, which is completely rural, is also densely populated, with around 1,260 persons/km^2.

The density of the population in Pondicherry is also quite high, slightly higher than Lakshadweep. The returns of the earlier censuses show that up to the 1931 Census, Pondicherry was the most densely populated area in the country which position is now being occupied by Chandigarh. Arunachal Pradesh is the most thinly populated area in the country. A district-wise analysis shows that those districts with high density fall in one belt which starts from the plains of Punjab, runs along the Ganga Valley, covers the east coast, goes round the southern tip of the country and covers the west coast. India's population is thus heavily concentrated in areas which have good irrigation facilities.

Some agricultural districts like Patna, Siwan, Vaishali and Dharbhanga in Bihar; Kanyakumari in Tamil Nadu; and Kozhikode, Trichur, Ernakulam, Alleppey and Trivandrum in Kerala have densities of above 800 persons/km^2. Besides, States and Union Territories where the availability of agricultural land is less, the density of the population is very low. For example, the north-eastern region of the country, which is mostly a hilly tract, has a density of less than 50 persons/km^2.

The first population census in India in modern times was taken in 1872. However, this census was not taken in all the States at the same time. It was only in 1881 that the first synchronous census was effected, and since then, such population censuses have been taken every ten years, the latest being that of 1981. Several scholars have tried to estimate the population of undivided India prior to 1872. It appears plausible to assume that the size of India's population did not increase significantly from ancient times till about 1600 A.D., the birth rate being only slightly higher than the death rate. In spite of this slow pace of growth, the population of India reached about, 255 million in 1871, equivalent to the population of the Soviet Union enumerated almost a hundred years later in 1970. During 1901 to 1981 there have been several territorial changes in the Indian domain and so the areas covered in the censuses have not been uniform. In 1936,

Burma was separated from India, and 1947 saw the partition of India into West Pakistan and East Pakistan. Again in 1951, the old princely states were merged with the Indian Union. The boundaries of a number of States were reorganised on a linguistic basis in 1956. Pondicherry, Goa, Daman, Diu, Dadra and Nagar Haveli were also merged into the Indian Union.

Subsequent to the 1961 Census, some boundary charges again took place, especially in the former States of Punjab and Assam and in the former Union Territory of Himachal Pradesh. Sikkim became a State of the Indian Union in 1975. Prior to 1921, the population of India grew at a slow rate. This was because India experienced a number of famines and epidemics. The famines of Mysore Madras, and Bombay, and took a heavy toll -of human life. The bubonic plague of 1896 resulted in about, one million deaths. The decade of 1901-1911 was more or less normal, but the 1911-21 decade experienced the great influenza epidemic in 1918 which took a toll of twelve million lives. In fact, during this decade the number of deaths exceeded the number of births which resulted in a decline of population.

After 1921, however, the movement of foodgrains became easier because of improvement in transport and communication facilities and as a result, famines were brought under control. Public health measures paved the way for a reduction in the intensity of communicable diseases resulting in the avoidance of national calamities. As a consequence, the population started growing at a faster rate.

The year 1921 is called the year of the "big divide" because during the 30 years prior to it. India's population increased by 12.2 million, or at an average rate of 0.4 million per year. But after 1921, India's population increased significantly, by roughly 28, 40, 42, 78 and 109 million during the 1921-31, 1931-41, 1941-51, 1951-61 and 1961-71 decades respectively. The growth of the population during the 1971-81 decade has been even higher, around 137 million. The rapid increase in population after 1921 is not due to any spurt in the birth rate but due to a decline in the death rate. For instance, during the 1881-91 decade the birth rate was 49 and death rate 41, and during the 1951-61 decade the birth rate was 42 and death rate 234 Control of epidemics and endemic diseases like malaria improvement in drinking water facilities, improved drainage system, increase in medical and health facilities, use of antibiotics and introduction of new drugs, development of a more effective food distribution system and avoidance of famine catastrophes of the past have led to a decline in death rate. The decline in the death rate was slow in the beginning but has been more rapid during the last two decades. From more than 40 per thousand population till 1921, the death rate declined to 36 and 31 in the 1921-31 and 1931-41 decades respectively and is, today, around 14 per thousand population.

The estimates of birth rates in India based on the registered data are gross under-estimates. However, by the use of the population figures of two consecutive censuses, it has been possible to obtain more reliable estimates of birth and death rates for the intercensal periods. It has been estimated that the birth rate in India in the early decades of the present century was very high and after 1921 it started declining. During the decade 1961-71 it declined to the level of 41.2. It was estimated to be slightly more than 41 in 1961 and perhaps remained at the same level till 1966.5 As a result of the family planning programme, which acquired momentum after 1965, about 6.8 million births were averted up to the year 1970-71. Again, another 27.5 million births were averted during the period 1971-72 to 1978-79 due to the adoption of various family planning methods in the country. As a result the birth rate in the 1970s further declined. According to the latest estimate the birth rate in India during 1971-81 came down to the level of 37.2 per thousand population.

The death rate in India is still considered to be high when compared to many of the Asian and Western countries. It is expected that the death rate in India will continue to decline and was believed to be around 13 at the end of the last decade. The decline is largely due to control of various diseases. But a further decline may be difficult unless the nutritive contents of the diet of an average person increase. As a result of the decline in the death rate, the expectation of life of India's population has increased. While in the 1921-31 decade the life expectancy at birth was 27 years, today it is estimated to be around 53 years.

In some of the Asian and most of the Western countries less than 45 children out of every 1,000 children born die in the first year of life; the number of such deaths in India is about three times higher. Out of these deaths which occur in the first year of life, nearly 85 per cent are in the first six months about 30 per cent occur in the first week; and approximately 53 per cent die in the first four weeks. Deaths in infancy are generally due to causes like tetanus, prematurity, dysentery, influenza, malaria, typhoid, etc. Female infant mortality in India is higher than the male infant mortality. Also, infant deaths are higher when mothers are young or comparatively older, when maternity takes place repeatedly and in quick succession. Reliable figures on maternal mortality in India are not available, but it is estimated to be around 10 per 1,000 live-births which is nearly 30 times the rate prevailing in the United States and other developed countries.

One of the consequences of a high birth rate is that India's population has an age structure which is typical of the countries of less developed regions-a very broad base and a tapering top. In demographic terms such an age structure is known as "young population". Nearly 40 per cent of India's population, according in the 1981 Census, is below 15 years of age and only 4 per cent is aged 65 years and above. Thus, 46 per cent of India's population depends on 54 per cent of the population in the economically active ages between 15 and 65, indicating that the "dependency ratio" is about 85 per 100 persons of ages 15-64.

The corresponding rate in the more developed regions of the world is about 55 only. A larger percentage of the dependent population tends to reduce savings and investments, and inhibits the rate of economic and social advancement since a large proportion of the scarce resources are diverted towards consumption. Also, an increasingly larger number of persons continue to enter the working ages swelling the ranks of the unemployed. Before the beginning of the First Five-Year Plan an estimated three and a half million persons were unemployed in India. At the end of the Third Five-Year Plan, their number increased to ten million even though thirty-one million additional jobs were created during the three plan periods.

The present age structure is also conducive to higher population growth as bigger cohorts of males and females continue entering reproductive ages, especially since universal and early marriage is the pattern in India. This would tend to further aggravate the unemployment problem.

The age structure of India's population for the 1981 Census, when compared with that of the 1971 Census, reveals that it has changed slightly and has become "older". The change is noticed specially in the age groups 0-9, 10- 19 and 45 years and above. In the age group 0-9, the proportion declined due to the decline in the level of birth rate. The quantum of decline is higher in the age group 0-4 compared to that in the age group 5-9, which suggests that fertility decline may be greater in the second half of the last decade compared to that in the first half. In the age groups 10-19 and 45 years and above, the proportions have increased which indicates general improvement in the health conditions, specially in the young and old ages.

A very large proportion of India's population lives in rural areas. The urban population in India in 1901 constituted only 10.8 per cent of the total population which in 1981 had reached 23.3 per cent. It increased very slowly from around 11 per cent in 1901 to about 14 per cent in 1941, but the decade 1941-51 showed an unprecedented growth from 13.9 per cent in 1941 to 17.6 per cent in 1951. In fact, during this decade the urban population increased by 41.4 per cent. This was partly due to the influx of refugees from Pakistan occasioned by the partition of the country in 1947. During the decade 1951-61, the growth of the urban population compared to that of the previous decade decreased which was mainly due to declassification of a large number of towns in the 1961 Census. The growth again picked up momentum during the last two decades. Roughly 86 per cent of India's 156 million urban population enumerated in the 1981 Census was from cities and towns having a population of 20,000 and more, and around 60 per cent or 94 million persons were from cities having a population of 100,000 and more.

There are more males than females in India. This is mainly because of higher mortality experienced by the females than the males. The trend of the sex ratio, defined as the number of females per 1,000 males, shows that up to the 1971 Census it was declining. However, in the 1981 Census it had slightly increased. The cause for the declining trend is attributed to the widening in the gap between the life expectancies of males and females. During the decade 1971-81, it appears that the death rate of females declined relatively more than that of the males. Union Territories of India for the 1971 and 1981 Censuses. In Kerala, there are more females than males. In Himachal Pradesh, Pondicherry, Orissa and Goa, Daman and Diu, the sex ratio is close to "balanced". However, in Uttar Pradesh, Jammu and Kashmir, Punjab, Haryana, Arunachal Pradesh, Nagaland, Delhi, Chandigarh and Andaman and Nicobar Islands the sex ratio is very low, lower than 900. In these States and Union Territories, apart from the differences in the mortality levels by sex, sex selective migration has also played a significant role in keeping the sex ratio at a low level.

Literacy, which is defined as the ability to read and write with understanding, increased slowly in undivided India during the period 1891 to 1931. But thereafter, it increased at a faster rate. In the Indian Union it reached 16.7 per cent in 1951, 24.0 per cent in 1961, 29.5 per cent in 1971 and 36.2 per cent in 1981. Literacy, and more so female literacy, is still low in India. However, the figures suggest that there has been considerable increase in the literacy rate and the gap between the male and female rates is narrowing.

A state-wise comparison of literacy rates shows that Kerala has the highest literacy rate 70.4 per cent, followed by Maharashtra with 47.2 per cent and Tamil Nadu with 46.8 per cent. The lowest literacy rate is observed in Rajasthan where it works out to less than 25 per cent. Among the Union Territories, except Arunachal Pradesh and Dadra and Nagar Haveli, the literacy rate in all the Union Territories was found to be more than 50 per cent.

In Arunachal Pradesh and Dadra and Nagar Haveli it was less than 30 per cent. Working Population Nearly one-third of the total, population in 1981 was returned as "workers". In the 1971 Census the work participation was at the same level but slightly lower, and in the 1961 Census it was around 43.

The reduction in the work participation rates between the 1961 and 1971 Censuses was mainly due to a change in the definition of workers. The Census Commissioner of the 1971 Census was of the opinion that the concept of "worker" used in the 1961 Census tended to inflate the number of workers because persons who are basically non-workers like housewives, students, etc. were included in the category of workers even though the participation was very marginal. In the 1981 Census, about 3.3 per cent of the population was returned as marginal workers.

Again, the same levels of work participation rates in the 1971 and 1981 Censuses indicate that during the last decade 1971-81, though a large number of jobs were created, they were not sufficient to scrap the unemployment situation. In the 1981 Census, about 51.6. per cent of the males and 14.0 per cent of the females were returned as main "workers". The corresponding figures in 1971 were 52.5 and 11.9 per cent respectively. In the rural areas 34.8 per cent of the population was returned as main workers and in the urban areas only 29.2 per cent was reported as main workers. The difference is striking in the case of females where their participation in the work force in the urban areas was very low.

The proportion of marginal workers is significantly higher in the rural areas than in the urban areas. Again, the industrial categories of the main workers reveal that a little more than 50 per cent of the workers in the rural areas are cultivators and more than 80 per cent of the workers are engaged in agricultural activities. The corresponding figures for the urban areas come to about 5 and nearly 11 per cent respectively. In the urban areas about 30 per cent of the workers are engaged in manufacturing, processing, servicing and repairs. About another one-fourth of the main workers are engaged in "other services".

India's fertility is higher as compared to that of the developed countries, but is relatively lower than that of other developing countries. It is higher than that of the developed countries because of universality of marriage, lower age at marriage, limited use of contraceptives, low level of literacy, poor level of living of the masses and the traditional way of life among 80 per cent of the population residing in the rural areas. It is lower than that of other developing countries because of a high incidence of widowhood, negligible number of widow remarriages, avoidance of coitus for a long period of time after childbirth and during religious periods, and the longer duration of lactation amenorrhea.

In India, because of early marriage, a woman tends to have her first child at an early age and continues to be highly fertile during the first half of her reproductive period. For example, it is not unusual for an Indian woman to have her first child before she reaches the age of 20 and to have three to four additional children before she is 35 years of age. The average number of children born to a couple who remain married throughout the reproductive period is between 5 and 6. This average number, however, is reduced to 4 or 5 when all ever-married women, that is, all who are currently married, widowed, divorced or separated are considered.

The pattern of age-specific fertility among females between the ages 15 and 45 shows that highest fertility is reached around age 25. While some data indicate that the highest fertility is reached between ages 20 and 25, some others bring out that the peak is reached between 25 and 30. It is not possible to resolve this controversy in view of the insufficiency of data. It may, however, be noted that fertility among Indian women remains high throughout the period 20-30 years of age.

Broadly speaking, the age-specific fertility curve shows a low but significant rate during the years 15-20, a sharp rise in the group 20-25 years with the peak being reached between 25 and 30 years. In the ages 30-35 there is a significant rate but somewhat lower than that found in the earlier age-group. After 35 the rate decreases appreciably, becoming insignificant after 40 years of age. Recent estimates of fertility rates made by the Sample Registration System have shown that fertility in India has declined. Reduction in fertility has been observed in all the age groups. The age-specific fertility rates bring out that the Indian fertility pattern continues to be of the "broad peak" type.

The age-specific death rates for all India estimated by the Sample Registration System for the years 1971, 1972 and 1976. The figures do not show any definite pattern of decline in the death rates. The 1976 figures in comparison to the 1971 figures show that the death rates in childhood and old ages have slightly increased. The reasons for the fluctuations are not clear and could be due to better reporting.

The recent estimates of the infant mortality rate obtained from the Sample Registration System show that it has reached the level of 130 and has remained more or less at the same level for a long time. In the urban areas the level of infant mortality is almost half of that observed in the rural areas. Again, by sex it is higher for females than males which is the reverse of the international experience. The expectation of life at birth based on the age-specific death rate observed by the Sample Registration System indicates that it continues to be favourable to the males as compared to the females. The difference between the two comes to around two or three years. Looking at the level of the death rate and the expectation of life at birth, it can be concluded that the present level of expectation of life at birth for India may be around fifty-three years which is close to the level assumed by the Expert Committee on Population Projection.

In the census the marital status of every individual is recorded but at the tabulation stage, since the 1961 Census, all those who are below 10 years of age are returned as "never married" irrespective of their actual marital status. According to the returns of the 1981 Census, about 50 per cent of the population have been returned as "never married" whereas little less than 44 per cent as "married". The widowed group constitute nearly 5 per cent of the population and a very negligible proportion is in the category of "separated and divorced".

Among the "never married" the proportion of males is higher than the females because males marry later than females. The same is also revealed by the proportion of "married" group by sex. Again, the proportion of widows is much higher than the proportion of widowers which may be due to the fact that more widowers remarry than widows.

Calculations based on the returns of the 1981 Census bring out that the average age at marriage of the female as 18.3 years; which is 1.1 years higher than that estimated from the 1971 Census. For males it works out to be 23.3 years. The mean age at widowhood of those who are widowed by the age 45 is 34.9 years and the mean length of fertile union for the decade 1961-71 works out to be 25.4 years.

India is one of the few countries where the age at marriage is still low. This is largely because of the prevalence of child marriages. Although child marriage was prevalent in many countries during the days of the Greeks and the Romans and in the early Middle Ages, the modern tendency is for marriage after eighteen years. Marriage in India is not, registered and therefore marriage registration data cannot be utilized to find out trends in the age at marriage. However, demographers have succeeded—through the use of census data on marital status—to calculate the mean age at marriage in India. The estimates show that the mean age at marriage of females which was 13.1 years in 1901 increased to 18.3 years in 1981. Similarly, for males which was 20.0 years in 1901 increased to 23.3 years in 1981.

The female marriage age has increased by 5.2 years over the period of 80 years during 1901-8 1, that is, on the average by 0.65 year per decade. The male marriage age, on the other hand, during the same period has increased by 3.3 years, that is, by slightly more than 0.4 year per decade. The increase in the marriage age is mainly because of reduction in child marriage pattern. For example, while in 1901, about 45 per cent of the girls in the age group 1044 were married, only 7 per cent of this age group got married in 198 1. Again, in the age group 15-19, while in 1901, about 84 per cent of girls were married, the corresponding proportion in 1981 was 44 per cent.

It was found in the 1911 Census that 17 per cent of the females were widows. The percentage of widows declined to 15.5 in 1931 and to 8.1 in 1981.14 In India all widows are not allowed to remarry as there are some social restrictions. The restrictions differ from caste to caste, but are most severe—nearing total ban—amongst the highest caste, and least amongst the lowest, the scheduled castes. The situation among other castes is mixed.

Due to the restrictions on remarriage, some widows are compelled to spend their fertile period in forced unreproductivity. Though this is socially undesirable, its effect is demographically favourable as it tends to depress fertility and lower the birth rate. Computations made with the help of census data on marital status by age reveal that the mean age at widowhood of those widowed by the age of 45 was 34 years in the 1951-61 decade, whereas it was around 33 years in the 1941-51 decade and around 30 years in the 1911-21 and around 29 years in the 1931-41 decades. For the decade 1971-81 it works out to be 34.2 years. A decline in the mean widowhood age during the 1911-21 decade could be because of the influenza epidemic which followed World War 1, and in the 1931-41 decade because of a larger number of child widows following a great rush for child marriages on the eve of the enactment of the Child Marriage Restraint Act in 1929. The recent increase in the mean widowhood age is attributable to improvement in mortality.

All castes in India do not impose restrictions on widow remarriages. According to some studies carried out in rural India, it was found that about one-fourth of those who are widowed get remarried. Of those who are widowed before the age of 20 and have no children, between 80 and 90 per cent are remarried but very few of those widowed after the age of 30 marry again.

It has been found that because of widowhood, a married woman has an average of one-half to one child less during her reproductive period. For example, it was found in Kerala that while females with unbroken marriages had 6.6 children by the time they completed their reproductive period, those widowed had only 5.5 children. In Madhya Pradesh the difference was only 0.5 child and in Karnataka one child in the urban areas and 0.5 child in the rural areas. India's population is composed of people of different ethnic composition, speaking different languages and believing in different religions. In this section the distribution of the scheduled castes and tribes, various languages and different religious are discussed.

In order to protect the interest of the backward classes and to uplift their social, economic and political status, the President of India is empowered to declare, through public notification, any caste or tribe or part thereof as Scheduled Caste or Tribe, as the case may be for the purpose of special benefit at is, about 15.7 per cent of the total population, were returned as members of Scheduled Castes. The number of Scheduled Tribes was around 52 million which constituted nearly 7.8 per cent of the total population. Thus, the Scheduled Castes and Tribes constitute nearly 23.5 per cent of the total population of the country.

Among the Scheduled Castes the largest number were enumerated in Uttar Pradesh where in the 1981 Census about 23.5 million persons were enumerated as members of Scheduled Castes. They are in sizeable proportions in Haryana, Himachal Pradesh, Karnataka, Punjab, Rajasthan, Tamil Nadu, Tripura, Uttar Pradesh, West Bengal, Delhi and Pondicherry where more than 15 per cent of the population has been returned as members of Scheduled Castes.

Though Scheduled Tribes are mostly concentrated in the far-eastern region of the country, the largest number were enumerated in Madhya Pradesh, about 12.0 million or about 23 per cent of the total Scheduled Tribe population. In Bastar district of Madhya Pradesh alone more than one million persons were returned as members of Scheduled Tribes. Their concentration shows that in Meghalaya, Nagaland, Arunachal Pradesh, Dadra and Nagar Haveli, Lakshadweep and Mizoram, more than three-fourths of the total population is tribal. In other states like Gujarat, Manipur, Orissa, Rajasthan, Sikkim and Tripura also they constitute more than 10 per cent of the total population. In India there are a very large number of languages spoken and in the 1971 Census more than 3,000 languages/mother tongues were recorded.

On scrutiny a number of them turned out to be cases of misreporting, different names for the same language, caste names, place names, names of religious faiths, etc. In the census the distribution of the speakers is given in two parts. The first part contains 15 languages which are specified in the Eighth Schedule of the Constitution. In the second part the distribution is given for 90 languages/mother tongues other than those mentioned in the first part. Among the languages specified in the Eighth Schedule of the Constitution, Hindi has been returned as the language/mother tongue spoken by a large proportion of the population. Other languages like Bengali, Marathi, Tamil, Telugu and Urdu are also spoken by a sizeable proportion of the population. Sanskrit, which is one of the languages mentioned in the Eighth Schedule, is spoken by a very small proportion of the population.

Since the 1951 Census, the office of the Registrar General of India has started publishing the total number of persons under different religious, groups. It was only up to the 1931 Census that cross classification of different religious groups by certain characteristics was available. The percentage distribution of the population by various religions shows that Hindu constitute more than 80 per cent of the total population of India. Muslims rank second with a little above 11 percent, Christians come next, forming nearly 2.6 per cent of the total population. Other religious groups like Sikhs, Jains, Buddhists, etc., are in the minority. According to the 1971 Census, though Hindus constitute more than 80 per cent of the total population of India, they are in a majority in only eighteen out of a total of twenty-two States in India. The States where Hindus are in a minority are Punjab, Jammu & Kashmir, Nagaland and Meghalaya. In Jammu & Kashmir, Muslims constitute nearly 65.85 per cent of the population while Meghalaya and Nagaland are dominated numerically by Christians. In Punjab, Sikhs form the majority.

Among the Buddhists, the largest number is concentrated in Maharashtra which accounts for nearly 84 per cent of the total Buddhist population in India. Jains, like Buddhists, numerically account for less than one per cent of the total population, though unlike them, they are spread all over the country especially in Gujarat, Madhya Pradesh, Maharashtra, Karnataka, Rajasthan and Uttar Pradesh. The population of India, according to some demographers, is likely to reach one billion before the end of the present century. So far India's population has grown because of a rapid fall in the death rate. In future it will depend more on the fall in the birth rate. The death rate which at present is around 14 is likely to fall to about 10 by 1991 or so.

Further decline will be more difficult and is likely to take a longer time. The critical factor in the growth of the population will, therefore, be the birth rate. It may be mentioned that the present long-term demographic goal of the population policy, which is laid down in the Sixth Five Year Plan document is to bring down the net reproduction rate to unity by 1995 for the country as a whole.

Development is frequently assumed essentially an economic concept and it has meant that capacity of a national economy, whose initial condition has been more or less static for a long time, to generate and sustain an annual increase in its gross national product (GNP) at rates between 5 to 7 per cent. For example, the United Nations accepted 1960s as the First Development Decade, in which development was conceived largely in terms of the attainment of a 6 per cent annual growth rate of GNP. The thinking of this period focussed mainly on the concept of successive stages of economic growth. Its advocate was W.W. Rostow. According to this doctrine, the process of economic development was seen as a series of successive five stages through which all countries must pass.

It was primarily an economic theory of development in which the right quantity and mixture of saving, investment and foreign aid were all that were necessary to enable less developed countries to proceed along an economic growth path which historically and been followed by the more developed countries. Therefore, focus was on rapid industrialization at the expanse of agricultural development. During this period, most common measure of development has been per capita GNP. In this context definition of less developed countries offered by Samuelson that "A less developed country is simply one with real per capita income that is low relative to the present-day per capita incomes of such nations as Canada, the United States, Great Britain and Western Europe generally". According to this criterian, roughly two-thirds of the world's people live in the 'low developed countries' with only about an eighth of the world's income.

Only 20 per cent of total population is in highly developed countries which jointly share about 60 per cent of world income. Asia with over half of the world's people, accounts for only 14.3% of production, while Africa and Latin America together have 16.6% of total population but produce only 6.5% of all GNP. Thus, the development was the state of affairs prevailing in the northern hemisphere and under-development were the circumstances existing in the southern one.

The western countries served as model of development to be copied by the developing countries also. The gap between richest (USA) and poorest (Rwanda) is more than 90 times. Another common economic measure of develop-ment is the growth rate of per capita GNP. It is pertinent to note that level and growth rate of per capita GNP do not covary. Most rapidly growing economies include high GNP countries, such as Japan, Israel and France and also some of the poorer ones— the Republic Korea, Papua-New Guinea and Nauritania. These two criteria yield quite contrasting patterns of development. Thus, whatever the development may be, it cannot be simultaneously measured by both the level and the growth of GNP per capita. In fact, GNP is incapable of presenting the real state of development, because of several short-comings.

Most of the countries particularly falling in the realm of the developing world, do not have evolved accurate accounting system, resulting in inaccurate GNP for them. At the same time, many of goods and services do not enter to cash economy. There is wide difference in methods and value systems of the countries of the world. Because of these technical reasons GNP of countries with contrasting level of development is not comparable. Further, it does not say anything about the level of human well-being, which is the basic objective of the development. It tells about the resources provided but not how they affect people's lives. To obtain the complete picture of development it is not sufficient to realize the amount of resources brought about by economic growth.

It is also necessary to examine the impact of these resources on the life of the people. Now a days, 'development' has become an alluring word and carries such a wide as well as specific meaning that common man as well as specialist both can use it in their own contexts.

In general usage, this word has been used in the sense of growth, meaning expansion in number or quantity of the same or collection of goods valued high. It has, most often, been preceded by some adjectives, such as economic, social, regional, industrial, agricultural etc. These contexts restrict the meaning of the development to a certain specific sphere of human domain. Later, it was realised that all spheres of human domain are intricately interrelated and one sphere cannot be developed isolating from the whole. Because of this very fact sectoral and target approaches of the development could not yield desired progress, despite heavy investment.

Consequently, 'development' is used as independent concept connoting the holistic approach of human progress, and every nation—developed or developing are striving for it.

Development is an old theme, but has been evolving new contents and fresh concerns. The realization of unacceptability low living standards among national populations as a whole is certainly the youngest, yet it can be traced back into the eighteenth century. The view of the development as the realization of the economic potentials is far more ancient, and the germ of the concept of economic and social progress goes back to classical time. Further, the concept of development has undergone serious evaluation and fundamental change over the past four decades. These changes were in part a response to the short-comings and failures of policies and programmes implemented during the 1940s and 1950s, and in part a reaction to changing economic and social conditions during 1960s and 1970s. The changes in perception of, and approaches to development can be traced by reviewing development theories and strategies during three recent periods of evolution.

The first period spread over 1940s and 1950s when international assistance organisations aimed at rebuilding the physical and industrial structure of countries that had attained relatively high levels of productive capacity before the Second World War and at re-establishing market mechanism in European economies. With the success in rebuilding European economies attention was turned to poor nations of the world and largely those strategies were blindly copied in poor countries which were used successfully in Europe. They aimed at maximizing economic growth, industrial output and export production through capital-intensive investment and were relying on the free operation of market system and 'trickle down' effects to generate sustained economic growth. It was assumed that 'industria-lization is the panacea for the economic progress, which will necessarily lead to social progress'.

The second period began by the early 1960s when it became increasingly apparent that in most developing countries the strategy of rapid growth through capital intensive industrialization was not working satisfactorily, and as such, during the period from early 1900s to the early 1970s, international development agencies and national governments were concerned with removing 'obstacles' or 'bottlenecks' to development. They primarily concentrated investment in 'key sectors' of the national economy and attempted to modernizing developing societies by transferring methods and industrialization from western industrial countries.

During the third period from the early 1970s to the present, international assistance organisations and developing country governments have been seeking to achieve more balanced 'development in the international economy by reducing the growing disparities between rich and poor countries, to achieve greater equity in the distribution of the benefits of economic growth within developing nations, to reduce the high and growing levels of absolute poverty and to provide for the basic human needs of those living in absolute poverty while increasing their productivity and income. In consonance with these changes the meaning of 'development' has been changing. Initially, it was synonym to economic growth. But this concept of development as economic growth is criticised bitterly for its short-comings and equity with growth is recognised as the objective of development. It is not a purely economic phenomenon. It is less as state than the process and is perceived as multi-dimensional process involving the reorganisation and re-orientation of entire economic and social system. It also involves radical changes in institutional, social and administrative structures, as well as in popular attitudes and sometimes customs and beliefs.

The experience of 1950s and 1960s led to the 'dethronement of GNP'. In fact, a large number of developing countries achieved the overall U.N. growth targets, but the level of living of the masses of the people remained for the most part unchanged. It means there was a gross mistake in the definition of development. Some theorists began questioning the assumptions underlying capital intensive industria-lization models. Dudley Seers, for instance, pointed out that the prevailing conditions that made rapid and sustained industrialization possible in western societies constituted a 'special case' and that economic development policies, applied successfully under those conditions could not be transferred or replicated in developing nations. He noted that in Europe and America factors of production had been abundant during the period of industrialization. But in developing nations the conditions were far different.

Attention was diverted to solve obstacles bottlenecks to industrial expansion during this period. As such, economic development was redefined in terms of the reduction or elimination of poverty inequality and unemployment within the context of a growing economy. 'Redistribution from growth' became a common slogan. Greater attention was given to programmes for asset redistribution, institution building, population and family planning, labour intensive and small scale industrialization, agricultural expansion, more appropriate use and conservation of natural resources and human resource development.

Key sectors of the economy were identified into which national resources and international assistance were channelled. Research into new high-yielding variety of seeds, construction of irrigation system, improvement of agricultural training and extension programmes, creation of marketing system, the organisation of cooperatives and credit schemes and land reform were major activities which were provided financial and technical assistance to strengthen the agricultural sector. Similar approaches were adopted in other sectors also.

Because of these targeted efforts, average annual growth rate in GNP was relatively high, but which simultaneously showed no improvement or witnessed actual decline in employment, equality and real incomes of the bottom 40% of their population. The emphases on exploiting natural resource on expanding industry rapidly and on concentrating investment in the large metropolitan centres of developing countries created increasingly severe environmental hazards and led to destruction or deteriorating of those countries' natural resource base. Now, the concept of development extends beyond the economic growth and quantitative measurements of income, employment and inequality. It is even more than that. The United Nations Research Institute for Social Development initiated work in this respect which was synthesized by Drewnowski. Drewnowski rejected such partial concept as 'economic' and social' development and has stressed the essential unity of the development process. He says, "Development is a process of qualitative change and quantitative growth of the social and economic reality which we call either society or economy.

The close interrelationship between economic and social elements precludes any purely social or economic development. It is, therefore, better not to speak of social development or economic development, but of a single process called simply "development". He regards the development as unfolding of creative possibilities inherent in a society" as opposed to growth which is merely 'an expansion of the system in one or more dimensions without change in its structure." Growth means a large cake, without such reference to its ingredients and with no reference to who gets now big a slice. The increased production is seldom, if ever, equality distributed or its composition equally approved.

The gist of the arguments presented at present is that the development is synonymous with progress, as experienced and evaluated by the people involved. Development is a welfare improvement, and therefore, those conditions and circumstances capable of enriching the welfare of the people are development. Welfare cannot be expressed by the per capita income. It is a subjective and value-based concept. It would be apparent by now that the meaning of development is very much a question of values, whether it is viewed as a complex, multi-dimensional phenomenon or merely equated with GNP per capita. As Bastar puts it, "Development is necessarily a normative concept, and involves values, goals and standards which make it possible to compare a present state against a preferred one. This raises immediately the question of whose values and goals are to be taken into account in assessing development. Planners' views or people's values?

Most national welfare functions, as far as they are explicitly stated, are a mixture of these different elements". In essence, development is improvement or human progress. It is reutterated by Goulet that "Development can be properly assessed only in terms of the total human needs, values and standards of the good life and the good society perceived by the every societies undergoing change. Although development implies economic, political and cultural transformations, these are not end in themselves, but indispensable means for enriching the quality of human life."

The survey of lessions of the last three decades, and reflections on the meaning and content of development leads to the following conclusions. First, development is more than material progress, it contains cultural, social, political and, above all, human dimensions. Second, development is largely an autonomous process and can be engineered from above only to a limited extent. The purpose of' this process is (i) to achieve a sustained increase in production, (ii) to distribute resources equitably, (iii) to respect the natural environment and the ecological balance and (iv) to modernize in a way that builds on society's indigenous values and meets the non-material needs of people. Third, no model of development can achieve these four objectives simultaneously and in all countries. Growth may conflict with redistribution, both way conflict with environmental concerns and conservation, and change and disruption may conflict with traditional values. Fourth, it is very difficult, if not impossible to follow a strategy of 'grow first and redistribute later'.

Finally the diversity of countries and people suggests that there is a need not for a single development strategy but for a range to suit the economic, social, political and cultural circumstances of different societies. Development is accepted as the sustained elevation of an entire society and social system towards a better or more human life. In this context, three basic components or core values, as suggested by Goulet, should serve as a conceptual basis and practical guideline for understanding the inner meaning of development. These core values are: life sustenance, self-esteem and freedom, representing common goals sought by all individuals and societies. While the necessities for sustenance of life in a purely biological sense are common to all people, those things on which esteem and freedom depend may be very much culturally specific. The meanings of these core values are broad and universal.

All people have certain basic needs to be fulfilled to sustain. These life sustaining needs include food, shelter, health and protection. A basic function of all economic activity is to provide people with the means of overcoming the helplessness and misery arising from a lack of food, shelter, health and protection. To this extent, economic development is necessary condition for the improvement in life; and rising income, elimination of absolute poverty, greater employment opportunity and lessening income inequality constitute necessary but not the only conditions for development.

Second goal of development is to create a sense of worth and self-respect. Such measures which provide identity, dignity, respect, honour or recognition to the members of the society lead to the development. For example, ownership of land is highest mark of social esteem and to perform manual labour the lowest in our country. Thus progress is associated with the land ownership, however, unremuncrative it may be. Therefore, change towards more equal distribution of land, or for more people to own land, would be considered progress in this sense.

Third and final objective of development is the freedom from the social servitudes of man to nature, ignorance, other men, misery, institutions and dogmatic beliefs. Freedom involves the expanded range of choices for society and their members, together with the minimization of external constraints in the persuit of some social goal which we call development. In this respect relationship between economic growth and freedom from servitude is significant, as wealth can enable man to gain greater control over nature and his physical environment. It also gives him the freedom to choose greater leisure, to have more goods and services. It can be concluded that 'development' is both a physical reality and a state of mind in which society, through some combination of social, economic and institutional processes, secure the means for a better life.

The components of better life vary from society to society but as identified by Todaro, development must have following three objectives in all societies:

— to raise levels of living, including, in addition to higher income, the provision of jobs, better education and more attention to cultural and humanistic values. These all serve not only to enhance material well-being but also to generate greater individual and material self-esteem; and

— to expand the range of economic and social choice to individuals and nations by freeing them from servitude and dependence not only in relation to other people and nations but also to the forces of ignorance and human misery.

The Society for International Development, Rome launched programmes of national and regional conferences with the objectives to explore thinkings on development problems and alternatives and to identity more meaningful strategies for future. The findings of these conferences are summarised by Richard Jolly. He asserts that both developed and developing countries are perceiving a point of change in world development in the 1990s.

In the developed countries, inflation and recession, energy crisis, terrorism and increasing sense of alienation and pointlessness, have increasingly been raising questions about the desirability and sustainability of past development patterns. After two hundred years, industrial revolution seems to be losing its confidence, if not momentum. Within developing countries, the perspectives are notably different. Poverty along with related problems of unemployment, rising prices and gross inequalities and imbalances, specially between rural and urban areas, is a continuing reality. There is also increasing awareness of the global dimensions of these national problems, and considerable unity over the injustices and asymmetries of the world economic system.

There is widespread agreement on the need for fundamental international changes if developing countries are to participate more fully and equitably within the world economy. The energy crises in the north, social restlessness in north and south and growing dissatisfaction with mass poverty in the south have become burning problems at present. In these perspectives, certain alternatives have been suggested to the existing conditions. These alternatives for future relate not just to changes in policy and action but much more fundamentally to changes in beliefs, values and objectives as to what should be achieved. This recent concept of development revolves around four key themes-basic needs, more growth, greater participation and self-reliance. These can form core objectives of development in rich as well as poor, industrialised as well as agricultural countries. However, the meaning given to these concepts and weightage given to them vary.

Satisfaction of basic needs are considered as foundation of development. A more appropriate definition of development would begin by identifying basic needs. As Paul Streeten suggests that in many underdeveloped countries the objective of development would be defined as raising the level of living of the masses of the people. This implies meeting such needs as continuous employment or secure livelihood, more and better schooling, better medical services, safe water at hand, cheap transport, and energy, and of course, a somewhat higher income.

The list of the conventional basic needs vary with the time, space and culture. However, they are mainly physical but also include non-physical needs, particularly in over-developed countries, where family and human relationship get squeezed by the pressure of industrial society.

Significance of economic growth and often acceleration of economic growth cannot be overlooked. In fact, any meaningful programme of meeting basic needs would require massive investment which would be possible by growth. Therefore, growth continues to be a priority development objective. Acceleration of growth will be needed in certain developed countries also to revitalise industry, to make possible more effective participation within the world economy and to permit a socio-economic change within the country. A development effort requires the full mobilisation and the direct and continuous participation of the people. This participation can be aroused by raising mass consciousness, decentralised decision-making and by encouraging collective participation in production, allocation and distribution of resources. It also depends upon the adoption of the technology which integrates human development with the local resources.

The fourth objective of development is that of, self-reliance. Self- reliance is used 'in the sense that the country has to manage its own problems to make its own decisions and to work for what it needs'. For assessing problems and potentials nation has to place reliance on the energy of the people. Thus man regains pivotal position in process of development both as agent and as beneficiary. It enhances development software: health as well-being, education as free access to knowledge and information, useful employment as a means of self-fulfillment. Development hardware, i.e. machinery, factories, vehicles, dams etc. is important because it helps in promoting the software and should be shaped accordingly. Thus self-reliance leads necessarily to reshaping the economy and society in order to meet basic needs.

The goal of people's participation was built into administrative organisation for developmental programmes, particularly for Community Develop-ments Schemes. Generally the objective of the public participation is to make the plan successful and achieve community goal. Another objective is to give planners full information about human needs and aspirations to mobilize pressure for prompt and effective action to represent more effectively the minorities.

Moreover, public participation enables people to share responsibility which lessens the burden of planners and make a plan realistic. Within a representative system of democracy public participation serves two major objectives: (i) Improve-ment in planning process by the dispersal and collection of information, (ii) Enhancement of citizenship by encouraging individuals or group of individual to take a more active part in discussion and determination of public policy.

The objectives of people's participation in developmental programmes and schemes can be conceived in four major tasks. (i) Plan formulation and decision making, (ii) Implementation of planned activities, (iii) Distribution of benefits of development, and (iv) Monitoring and evaluation. Participation of the people in institutions and systems which govern their lives, is a basic human right and also essential for realignment of political power in favour of disadvantaged group and for social and economic development. It is a group for social channel through which people ensure their effective influence on the decision making process at all levels of social activity and social institutions.

Developmental planning may fail if the cooperation, association and involvement of beneficiaries, particularly the poor sections of the society are not sought. In the third world countries, people are no more the subject of development, they have become object of development and in many cases even resources for development. For the planners and bureaucrats people are the vague target group. Since peoples active participation strengthens democracy and facilitates a vigilant watch on administration and helps in accelerating the pace of achieving the targets of area development programmes, the introduction of peoples participation technique in the planning process is must.

The main issues involved in the successful implementation of area development programmes in India are promoting the motivation of the people so as to secure their full participation. The Patha area (1165 kM2) which is a sharply defined upland plateau, lies in the south-eastern part of Banda district (U.P.) This most backward area of Bundelkhand is a drought-prone as well as one of the worst affected water scarce areas of U.P. because of its lithology, physiography and climatic conditions. Its 88% working population is employed in agriculture sector.

People in majority are marginal farmers and landless labours. Industrially it is a backward area in the presence of ample basic raw materials. The study area with a population of 80248 is scheduled castes dominated area where the kols have been the subject of worst form of exploitation by elites, and lead a life of helplessness. Poor literacy (15%), prevalence of bonded labour system, the rigid caste system, and the inherited administrative and bureaucratic system of the 'Raj', and the constraints of physical factors all have made the poor sections of the society as the object of development. Moreover, the study area suffers from poor rural-urban relations, very low density of roads, poor communication linkages and other social facilities.

Non-involvement of people in rural drinking water supply schemes seems to be one of the main factors not only for the insignificant impact of such schemes but also for their failure. There is a need to examine the strange and paradoxical situation of the rural drinking water supply schemes of the Patha area which failed to solve the problem of acute water shortage. The main objectives of the present study are to (i) examine the objectives of people's involvement in rural drinking water supply schemes, (ii) highlight the problems faced by the beneficiaries in involving themselves in the formulation and implementation of such schemes and (iii) suggest a strategy for people's participation in the successful implementation of such schemes under the existing socio-economic and administrative set-up of the study area. In the present study out of 106 inhabited villages of the Patha area only 23 have been selected on the basis of purposive random sample technique. 10% of the households were covered by interview schedules which included queries about the nature, quality and present condition of available drinking water; management and maintenance of water supply works, the involvement, acceptance and cooperation of beneficiaries to such schemes.

The Patha area is well known for extreme heat and scarcity of drinking water in the months of summer. Besides the acute shortage of water, the meagre water which is available, is brackish and contaminated. In order to supply safe and treated water in rural areas the then L.S.G.E. Department (Now Jal Nigam) started two schemes: (1) The Patha Rural Drinking Water Supply Scheme, (2) The Bargarh Drinking Water Supply Scheme. The former is one of the largest schemes in Asia which was executed in January 1974 with a total cost of Rs. 198.68 lakh. The Patha scheme covers 1254 km2 area and includes 105 villages and two towns, i.e. Karwi-Chitrakut and Manikpur but at present only 58 villages of the study are linked with it. Under this scheme 7 (large size) and 71 (small size) overhead tanks, 525 public stand posts have been connected by 570 km long pipe lines.

As per design this scheme is bound to supply 45 litres of water per unit daily in rural areas. The Bargarh Scheme includes 33 villages of Mau tahsil which lie south of Vindhyan range. This scheme costing Rs. 28.42 lakh, was executed in the year 1973-74 to benefit 14288 people. Under this scheme there are 9 overhead tanks and only one public stand post. This scheme is designed to supply 60 litres of water/unit daily.

Really it is a paradox that in the presence of huge Patha Drinking water supply scheme people are still facing acute shortage of drinking water. On account of non-cooperation of beneficiaries, village panchayats and very low number of private connections, Jai Nigam was incurring an annual loss of Rs. 10 lakh for the maintenance and other works of the scheme.

As the State Government was not interested to extend an annual grant of Rs. 10 lakh for this purpose, Jal Nigam stopped water supply on January 1, 1976 to those villages which had not taken private connection or failed to deposit water tax. Ninety one villages of the study area are included under these two schemes but the treated water is supplied to only 60 villages. At the time of survey only 1.5% of the households have had private connections because the water tax burden was heavy for poor farmers and kols. 62% of the stand posts are of no use as their water supply is not only irregular but also meagre in quantity. The fact is that people became bound to withdraw their connections. At present Jal Nigam has installed India Mark-11 handpumps as an alternative source for drinking water but the handpumps in majority are not in operation.

Water supply is an economic asset to the community, But the common tendency of consumers is to get as much water as they can, and to achieve it they do not hesitate to temper with the distribution system. Due to the sheer lack of the feelings of community welfare, the majority of the consumers understand that it is meant only for their own benefits. A large number of rural water supply schemes are running in loss and only a few of them are self-sustaining due to lack of public acceptance by way of unfavourable response to individual house connections. As participatory planning involves people at all the phases of planning, i.e. pre-action, action, and post- action, the beneficiaries should not only involve themselves in the decision, implementation and monitoring processes but also give full cooperation.

In the pre-action phase of these two schemes the beneficiaries and local bodies, such as village panchayats did not associate themselves in the formulation of the schemes. Truly speaking, presently, there is hardly any participation of the rural poor in any kind of decision making at local level. Normally, the policies and programmes are handed over from above and local people are required to fit in this framework. Since these two schemes of the study area were not pilot projects like 'Banki' rural drinking water supply project, people's participation was not encouraging, rather to say it was insignificant. During the action phase, the normal practice of people's participation in the implementation of developmental programmes has been in the form of construction of free labour, voluntary or otherwise. In the study area the beneficiaries have either voluntarily given their land or remained satisfied even at low compensation rate for the construction of overhead tanks and standposts. But they did not contribute in form of free labour in the construction works.

In case of drinking water supply schemes, people's involvement in post-action phase is must. The right index for measuring the acceptability of a scheme by the people can be found by their urge to maintain the facilities provided. In case of the rural drinking water supply schemes of the study area, majority of the beneficiaries did not take connections. The poor sections of the rural communities in the study area show a marked reluctance to pay for water as in rural areas alternative sources are almost always available except the months of summer.

Most of the gramsabhas also did not cooperate in sharing the nominal cost as required by them. The officials of Jai Nigam unusually complain about the misuse of the distribution system, particularly the damage to pipe lines and a general attitude of negligence towards the leakage from pipe liners.

Since monitoring and evaluation are essential to provide feed-back to the project management to improve the efficacy of the delivery of services and inputs, the leaders at local level have always complained about the mismanagement of the water supply system to concerning officials. The system of delivery of services in the entire range of direct social services sectors is controlled the existing administrative system. If we look at the planning process being followed in developing countries, we come to realize that it is essentially a bureaucratic and technocratic exercise. The plans are conceived by elites, documented by elites, and finally implemented by elites.

The bureaucracy failed to ensure the people's cooperation in C.D. programmes due to its rigid approach, conservatism and lack of development attitude. Moreover, bureaucracy has also rubbed a part of itself on voluntary organizations through grant-in-aid patronage and other subtle means. The representative democracy as pursued by the elites of the developing countries has not been able to provide vehicle for people's involvement in developmental programmes. Another reason for non-involvement is that they do not possess the requisite knowledge and expertise to participate in complex decision.

The main hurdles in the non-involvement and non-cooperation of beneficiaries in rural drinking water supply schemes of the study area are the rigid approach, conservatism and lack of developmental attitude of the bureaucrats and technocrats towards peoples participation. Due to lack of education the people showed a poor awareness to the importance of treated drinking water supply.

People of the study area, in majority, do not possess functional skill, such as reading and writing needed for effective participation in the formal meetings. Moreover, they also lack the urge to maintain this facility. Their illiteracy compels them to disassociate from the group discussions and even to learn the elementary principles of hygiene, and to lodge complaints against the irregularities of water supply and even the workings of the technicians and officials. Their poverty is also a great constraining factor. Their dependence on rural elites, ex-zamindars and moneylenders also compel them not to associate and involve themselves in the proper implementation of these schemes. The dependence of poor sections of the society on the rural elites their poverty and literacy have become a hindrance in the implementation of development programmes. These two schemes could not be successful as the majority of the beneficiaries who are poor, did not take private connection on account of their poor capacity to pay the water tax.

Development planning, especially at the grass-roots level, can realize its full potentials only if people are motivated and actively involved in an organized way. Similarly participatory planning presupposes planning by the people. Once the people are placed in the centre, the process of planning could be conceived in terms of triangle whose interior is covered by the peoples and at the vertices are the three participants, i.e. government, voluntary agencies, technical and financial agencies.

The participants assist the people; they do not perform the role of leaders. Thus, participatory planning process is one which has effective participation of the people as its major operational objective. It is a process by which individuals, lowest in the organizational scale, are enabled to have effective voice in decision making. The Sixth Plan Document, thus emphasising the role of people, states: "Democratic planning means of the harnessing of people's power and their fullest participation".

Since the objectives of the people's participation in rural drinking water supply schemes is to make the scheme successful and to achieve the community goal, there is a need to formulate a strategy for people's involvement. Patterns and techniques are the two principal issues which must be decided at the grass-root level keeping in view the political system and socio-economic set-up of the region. As regards the patterns of people's participation in drinking water supply schemes, some questions: (1) who should participate? (2) how much participation is feasible and desirable? (3) on what issues people's participation is desirable? and (4) what weight should be attached to their views?, must be answered. Some are of the opinion that the ideal patterns of people's participation must cover the hole population of a community. But making every citizen to participate in such schemes is not practical.

On the other hand if the elites of the rural community, the so called the representative of the people, are only involved, it is but natural that the interest of the poor majority will be over-looked. Participation is not some process in involving every one and rendering all common to denominator. Rather it consists in involving institution structures from which diverse individuals get a some of dignity and self-respect.

Since drinking water supply comes under the planning for infrastructure or environmental planning and involves technocrats, it is not feasible and desirable that all the community members should participate in decision making and formulation of the schemes.

Participation cannot be imposed on people from above, it should be voluntary and based on will to participate. People's involvement in developmental programmes can better be understood in four senses: (1) decision making, (2) implementation of projects and programmes, (3) monitoring and valuation, and (4) sharing the benefits of development. In the present case all these four aspects of people's participation have been covered under three phases of rural drinking water supply schemes, i.e. (i) Pre-action phase (ii) Action phase and (iii) Post-action phase. A model (Fig. 3) has been developed to depict the role of peoples participation in different phases rural drinking water supply schemes/ programmes on one hand and the role of government in seeking people's participation on the other hand. This model clearly points out the line of strategy for the peoples participation for the success of such schemes.

The feasibility and desirability of people's participation in such schemes depend also on the different phases of the planning. During the pre-action phase, i.e. the plan-formulation and decision-making phase, people should help in basic data collection. Data collection process through community volunteer is the first step towards operationalization of participatory planning. In the beginning the rich are likely to influence the decision in their favour in setting priorities of the projects and their location.

People must attend public meetings, participate in group discussions arranged on the various aspects of treated water supply. They should help the planners in the identification of the problems, its severity, its intensity and its magnitude. For the identification of these aspects women must be involved as they face the real problem of bringing water for drinking purpose from distant places, particularly in the scorching heat of summer months. A great deal of energy and time in particular of womenfolk are at present consumed in drawing and carrying small quantifies of water to home. During the action phase the beneficiaries must co-operate in the construction of water supply system in form of free labour voluntarily. The existing practice of people's participation, through free labour in both inequitable and exploitative of the poor.

An equitable participation is possible only when everybody in the village either voluntarily contribute or is required by the community to contribute on equitable share. They should also give their precious time in the protection and up-keep of the installations. The post-action phase is of great importance because it includes the implementation, sharing the benefits and monitoring of the scheme. During this phase which covers the longest period, not only the cooperation and association of the beneficiaries but also their involvement is most essential as without that the efficient and uninterrupted maintenance of the project is rather a difficult job, inspite of the strict vigilance, supervision, and careful working of maintenance organization.

In order to make the scheme successful people's cooperation has to be secured effectively and make them interested in the scheme, and conscious of the ownership and promotional participation. It is imperative on the part of beneficiaries to understand that these schemes are not meant for the benefit of any particular individual, but for the community as a whole. People's promotional participation is must to reduce the wastage of treated water, to take action against miscreants and to restrict use of water in drought period. Period of two decades has almost passed since the execution of these schemes, the beneficiaries must have understand the importance of treated water, they should cooperate with Jai Nigam by taking private connections and depositing the pending water tax.

There is also little participation by the majority and the poor in monitoring and evaluation of the schemes at present. To identify not only how many but also who benefits from a particular projects, people's involvement in monitoring and evaluation is also essential because it can reduce management of resources in a project. The beneficiaries should cooperate in identifying the related problems and constraints in the implementation. People should also evaluate the performance of the technicians and the government personnel attached to these two schemes. The role of government agencies or delivery systems is important one to supplement the efforts of the community through its mechanism. The bureaucratic attitude towards citizen, participation must be changed since India has adopted a democratic political system based on the competitive participation of citizen in the management of public affairs.

The existing system of the delivery of the rural drinking water supply schemes which is a part of the administrative and bureaucratic process, should be changed. In the formulation and decision making process, i.e. in the pre-action phase the government personnel should take the beneficiaries into confidence while deciding the goals and finalizing the tariff policy. Complementary investment should be made by the government so that the beneficiaries are not overburdened. Government personnel must have a socio-economic and behavioural survey of beneficiaries first and then through joint committees, close contacts with the people should be made to motivate them for accepting this technology. They should try to impress upon the people about the importance of treated water through group discussions, propaganda and other means. On the whole people must be educated to the advantage of treated water; and the protection and maintenance of the distribution system.

Motivation and incentives to accept these programmes should also be given by the government. They should involve people in data collection and in the identification of problems. During the action phase the beneficiaries should be involved in the protection of the establishment under construction and maintaining their function. During the post-action phase the role of government personnel for seeking the cooperation of beneficiaries becomes important because without their cooperation there is possibility for the failure of the scheme. As the people of the study area have no understanding of the basic principles of these schemes, they can be trained and educated to realize that potable water is essential for good health and it costs something to them.

The lack of response from the villagers for maintaining the water supply schemes is quite marked at the places. Thus it should be a part of the programme to make the villagers realize that these schemes are their and are being implemented for them. Government personnel should persuade the people to take private connection for treated water. The beneficiaries must be associated in the efficient maintenance of the distribution system. The right index for measuring the acceptability of the scheme by the people can be found by their urge to maintain the facilities provided to them. In the study area, Jai Nigam has installed India Mark-2/Mark-3 and pumps in almost all villages due to failure of these two schemes. During 1989 a government sponsored agency 'Mahila Samakhya' trained some women for repairing these handpumps. At present the trained women repair India Mark-3 handpumps after getting information from water members. Thus the women are looking after the maintenance of the said handpumps, the main source of drinking water in this water-scarce area.

For the effective implementation of these schemes the government personnel must have sympathetic attitude towards beneficiaries in solving their problems related to water supply, billing and other associated grievances otherwise they will loose their cooperation. The final step would be to work out in details the institutional framework which would be needed to implement the programmes in an interested manner. A number of functional grounds and sub-committees should be constituted so that almost every household is involved in the process of development of this important infrastructure. But before finally arriving at a plan of action, the project proposals must be discussed by the villagers for a number of times.

The formation of bigger groups should be discouraged. The groups should not be multifunction in scope. However, die mobilisation of the poor is not an easy test. So the technique should aim at attracting this 'average members' and minorities in the formulation and effectuation of the plans. Among the methods, the representation of groups is considered as the best technique. Community group can act as a pressure group. The growth of community effect should be encouraged as a part of planning process, not through a middle bureaucracy or through patronage of grant-in-aid, but a legitimate and integral part of the system. Moreover, the success of systematic participatory planning and development management would, however, depend upon training and orientation of the official.

In the pre-Independence period the population was more or less stable with high mortality and fertility levels. Subsequently the decline in mortality, without any corresponding decline in fertility has resulted in accelerated population growth.

The annual exponential growth rate of population was as high as 2.28 per cent for the decade 1971-81. Since Independence there has been a near doubling of the country's population. At present the population is near 800 million. The rural population as per the 1981 census was 76.7 per cent and the urban 23.3 per cent. Expectancy of life has been gradually improving from 32 years in 1951 to 58.6 years in 1987 but, in comparison to advanced countries, life expectancy in India is still quite low. The age pyramid of India's population shows a broad base with the dependency ratio as high as 85 in 1981. It is also estimated that the young age structure holds the potential for rapid population growth In future years unless specific measures are taken to reduce fertility. The accelerated growth of population has been a matter of great concern for the government.

A National Family Planning Programme was introduced in the First Five Year Plan with a modest investment for contraceptive services in hospitals and clinics. From the Third Five Year Plan onwards, the programme was strengthened for improving the outreach of the services in the community with an extension education approach and has been given a high priority among the programmes.

The desire of the government was to bring down the birth rate to 25 per thousand at the end of the Fourth Five Year Plan. However, the acceptance of contraception has not been up to the desired level. Birth rate has been brought down only to 32.9 in 1985. The effective couple protection rate in 1986-87 was only 37.1 per cent. It has also been observed that the mean number of living children of sterilised cases was 3.5 and of IUD acceptors

2.3. It has been realised only now that for a meaningful check of the population, contraception should be practised by much younger couples for spacing of children. We shall now consider the issues under development. The aim of human development is to improve the quality of life which depends primarily on the fulfilment and attainment of basic needs for survival and security.

The main thrust in all the Five Year Plans was increasing agricultural production for achieving self-sufficiency of food products. Next to agriculture, industry was given priority. Only at the end of the Fifth Plan agricultural production exceeded the target showing that there was surplus of food-grains. The objectives in all the plans included increase in national income, expansion of employment opportunities, reduction in income inequalities and eradication of poverty. It was only from the Sixth Plan onwards that a significant reduction in poverty ratio' could be noticed from 48.3 per cent in 1973-78 to 37.4 per cent in 1983-84. Let us review the situation to see to what extent the development programmes have helped the common man regarding the basic needs of food, shelter, water and fuel and also education which has generally been accepted as a basis for progress; and study it vis-a-vis population growth.

The available statistics about the production of foodgrains claim that after the Fifth Five Year Plan, the country has become self-sufficient in foodgrains and that there has been such a surplus production that it was possible even to tide over the crisis of drought in some states in 1987. However production is not a guarantee for the availability of food to the lowest economic segments of the population in the country.

An analysis of the poverty profile in India and its projection in the future years indicate that even by the year 2000 A.D., there will be a sizeable poverty stricken population in the country with very limited purchasing power. Public Food Distribution System and increased income generation for poverty alleviation are two important measures needing the greatest attention. Mahatma Gandhi was repeatedly stressing the need for self-sufficiency in the vilages not only for food but for all consumer items required in daily life. During the British rule, Gandhiji was strongly preaching "Swadeshi Movement" and boycott of foreign goods. What he said in the context of foreign rule still holds good when we look at the present relationship between the cities and villages.

In the case of all food items the production takes place in the rural areas but the rural folk, while they have to work hard to produce the food commodities, are not in a position to get enough for themselves; moreover their standard of life and the amenities they have are much below what the city dweller gets. There are many steps which the government could have taken from the time of Independence which would have ensured enough to eat for every individual citizen.

Land reforms are necessary in the first instance to give opportunity and impetus to small land holders to grow on their own soil and to avoid paid or bonded labour, indebtedness etc. Adequate irrigation is necessary to bring more and more land under cultivation, This is where the investments and efforts of the government have not been enough nor, appropriate. It is difficult to understand why large quantities of water should be allowed to go waste into the sea and in spite of repeated floods every year the government has not taken action to divert the water through a grid system of rivers, canals, drains etc., which had been recommended by veteran engineers like Sir M. Visweswarayya and Dr. K.L. Rao. Such plans would have helped food production, communication and traffic, prevention of floods and also avoided interstate disputes and above all provided employment for many. Huge sums of money spent every year on flood control measures and flood relief, could very well have been diverted to conservation of water. At present agriculture in the country depends on the vagaries of nature which are very difficult to predict and, therefore, there are fluctuations not only in food production but also in employment opportunities which have resulted in migration from villages to towns and cities as a continuing phenomenon.

There is no point in priding ourselves that India stands seventh among all nations in industry when it has not been possible to provide the basic needs of food and shelter to a sizeable proportion of the population. As Gandhi said, the poor of the world cannot be helped by mass production but by production by the masses. The system of mass production, based on sophisticated, highly capital intensive, high energy depending and human labour-saving technology is not only irrelevant but also encourages the rich men to extract work from poor men and widens the imbalance of wealth whereas the technology of production by the masses making use of available knowledge and experience is conducive to decentralisation, compatible with the laws of ecology, gentle in its use of scarce resources and designed to serve the human being instead of making him the servant of machines.

Alvin Toffler has clarified how, with rapid industrialisation and automation, the natural resources which cannot be renewed are very fast being depleted while there is always scope for living a modest and contented life by utilising renewable resources which will be available for ever. In his books *Future shock* and the *Third Wave,* he has warned the developing countries against lop-sided development with concentration in cities and towns at the neglect of villages. Schumacher has also made a similar point in his book *Small is Beautiful*.

Keeping in mind the recommended norms of space, light, ventilation, type of flooring etc., academic surveys and also the census data have brought out that about 60-70 per cent of the people live in substandard houses both in rural and urban areas. There is a glaring imbalance between the few rich and the vast poor. It is for serious consideration whether the situation is to be attributed mainly to the rapid population growth or whether a better pace of development in regard to housing could have been achieved in spite of the growing population. Housing like any other necessity of life is dependent on the earning capacity. But the next question to be asked is whether and how the government has approached this problem.

Investments of the government on housing schemes have been mostly in the urban areas and in general as yet no appropriate scheme has been undertaken in rural areas. The cities and towns are expanding out of proportion whereas material for construction is scarcely available in the rural areas. There is often a mention of appropriate technology for construction in rural areas but nothing has happened yet in practical shape. In the village it is only a few with financial resources who are able to live comfortably whereas the large majority have to huddle in huts and semipermanent structures.

While grandiose schemes for beautifying the cities with gigantic monumental structures, gardens, stadia, auditoria etc., are continuously being undertaken, there has been no interest shown by the government to invest uniformly and consistently for improving living conditions in the villages. It is because of the poor sanitation and poor living conditions that some of the major communicable diseases are persisting and difficult to tackle. Malaria and Filariasis could have been eradicated merely by proper drainage.

In advanced countries in the West diseases like poliomyelitis, diphtheria and tetanus were got rid of much before the discovery of vaccines and that was possible mainly because of improved sanitation and education. In olden days when man evolved from the hunting and food gathering to the agrarian society the tendency was to live near rivers and lakes. Later digging of wells was resorted to when the settlement was away from rivers. Scarcity of water for drinking and washing used to be talked about a few decades ago only in desert regions and in places where subsoil water was very deep. But with the rapid migration that has been taking place from rural to urban, there is no city or town in India today which is not facing a shortage of water. Unfortunately the situation is getting worse in the rural areas also because of neglect of old systems which were there to impound and store rain water.

In very recent years some steps have been taken by the government to provide bore-wells in all villages under the "Water Decade" scheme. Unless steps are taken to conserve water available from nature, and prevent its wasteful flow into the sea, it will be difficult to have a continuous source of' water supply either from the surface or from below. It is the proper management of water that is required and it will be futile to say that the shortage of water is due to population growth. Traditionally the main source of fuel in the villages was from trees and bushes. There was not much destruction of trees because only the dried leaves, twigs and branches which were being discarded by the trees were collected and used. Unfortunately it is the lopsided urbanisation and migration which are responsible for a fuel crisis that is now in evidence. Left to themselves villagers were not removing trees wholesale; they were only taking what the trees were shedding but the demand from the cities and towns for timber for construction and fuel in household as well as for industry and power generation has motivated rich and greedy contractors, to devastate trees and denude forests on an unprecedentedly large scale. This is being very much criticised by environmentalists but not much has been done to restore the natural equilibrium in the ecology.

One of the teachings of the Buddha was planting of trees; and in the days of Ashoka and other great rulers, tree plantation was observed and the whole area of India was covered with trees, free of dust, with plenty of water, plenty of shade, plenty of food and materials. It was realised by our forefathers and seers that God had given trees for almost all human needs, But the trees are having the same fate today as the goose that laid the golden egg. We cannot certainly say that there is a shortage of fuel due to a growth in population.

Education has been well recognised as a very important component of development. From the First Five Year Plan the intention of the government was to make education compulsory for every child and a system of free primary education has been in vogue. But in effect there has been no compulsion for education nor has the free education been appreciated and availed of adequately.

The literacy rate over the four decades has shown very little improvement and at present except for a few states in the country the literacy status is very low as compared to advanced countries (the literacy status was 16.7 per cent in 1951 and rose up to 36.2 per cent in 1981. One could straightaway say that the infrastructure was not adequate and it could not cope with the growing population. At the same time the number of children who attended school was so small that it could not be argued that the infrastructure was not enough. Moreover, over the past four decades, we have witnessed quite a huge number of private schools catering to the needs of those who could afford to pay. However, the governments have not done anything specific enough to ensure that every child gets some education formal or informal.

Moreover, the present system of schooling and collegiate education has only resulted in large unemployment and frustration and also made the situation -worse by taking the youth away from their traditional trades and leaving them in the lurch; and thereby compelling them to leave the villages. Two all-India surveys were conducted by the Operations Research Group, Baroda regarding family planning practices in India in 1970 and 1980. According to the first survey there was a fairly high awareness in all the states about the terminal methods of contraception but very poor awareness about the temporary methods, but at the same time the acceptance of contraception was only about 14 per cent.

The situation had only slightly improved in 1980. The awareness of all the methods has increased but is still very low in the rural areas, Regarding contraceptive practice the percentage of users has gone up to 35 in 1980; The ORG report has brought out, in addition, the following points:

1. More acceptance of family planning in urban areas than in rural
2. More acceptance of family planning - in higher castes of Hindus than in scheduled castes and tribes.
3. Education, particularly of wife, had a positive association with family planning.

4. Family income was also positively related with contraception.

5. Practice of temporary methods was more among educated groups and higher income groups.

The Family Planning Foundation, New Delhi had made a study in some selected states under the title "Diagnostic Study of Population Growth, Family Planning and Development 1971-81". The reports were published in 1985. From these studies it is evident that literacy, electricity and road communication and health services infrastructure had been found to have a positive association with family planning acceptance.

It can be inferred also from the state-wise statistics that developmental efforts have definitely helped in the promotion of contraception and the acceptance of a family size norms. In this context we may also refer to experiences in other countries establishing the positive influence of rural development on fertility behaviour. There is no need for any pessimism about the rising population offsetting the gains in any field of development. Decent and tangible progress has been made towards development; the only thing now required is proper prioritisation and shift to ensure uniform and accelerated growth of rural economy and slowing down wasteful and irrelevant expense on the urban side. With regard to family planning, the progress is admittedly slow and it is very glaring that enough awareness about family size limitation and available methods of contraception has not been created in the rural community.

Since enough evidence has accumulated to indicate a definite positive association of socio-economic development with fertility control there is need for a coordinated effort, to ensure that every adult is engaged in productive work and gainful occupation.

The main thing to be remembered for some years to come is that the village is the backbone of society and, therefore, the focus should be on the rural areas with highest priority. Needless to say agriculture has to be diversified and expanded. Some of the important steps to be taken will be:

1. Land reforms for equitable distribution of land and removal of bonded labour, indebtedness etc.
2. A country-wide inland water management for conservation of water, permanent irrigation facilities and conservation of soil.
3. Improved technology for cultivation.

In the case of industrial development the following points have to be borne in

1. It has to be more rural based than at present. The idea of establishing poles of development around towns and cities should be given up.
2. Stress should be on agro-based industries. 3. Production should depend more on human energy and skill than on machines.
3. Production by the masses rather than mass production.
4. Production should be need-based and relevant for daily consumption and should not fall into the category of luxuries.
5. Export should not become a consideration for commercial purpose till enough has been produced for local needs.
6. Promotion and preservation of local artisanship and craftsmanship.
7. Promotion of poultry farming, processing of animal foods, fisheries etc. at village level.

It is a very good augury that a separate Ministry has been formed recently at the centre for Human Resource Development. Quality of life depends largely on health which in turn is influenced by a variety of socio-economic factors and also accessibility of health care facility. Besides increasing the employment opportunities and earning capacity considerable attention has to be paid by the government in the rural areas to proper housing, water supply and sanitation.

A huge bulk of the communicable diseases are bound to disappear if sanitation is improved. Appropriate technology at low cost has been tried out successfully in various parts of the country and also outside for the safe disposal of human excreta. While there used to be considerable resistance among the rural folk to accept handflush, water seal latrines and bore hole latrines a few decades ago people have started asking for latrines because in growing villages open space is also diminishing. The sludge in the latrine pits has manurial value; and gas plants using human as well as cattle excreta are also gaining popularity in rural areas. Similarly soakage, pits and kitchen gardens have been found appropriate for disposal of waste water.

Time is ripe now for having a nation-wide rural reconstruction scheme with focus on housing, sanitation and safe water supply. Human labour for such work can be had locally and that would also be an incentive for gainful occupation. For safe water the recent experiences of UNICEF/WHC-assisted schemes of bore-wells with pumps have been quite cost effective and feasible; the chances of contamination and wastage have been minimal. This scheme has to be implemented widely and with urgency.

In this context, it may be mentioned that India is one of the member countries of WHO to have endorsed the resolution at the Alma Ata Conference in 1978 of 'Health for all by 2000 A.D'. and primary health care. It has been agreed that there should be a proper intersectoral coordination to ensure that the development programmes are so dovetailed into one another that there will be a synergistic effect for a quick pace of total human development. It has also to be remembered that sanitation and proper nutrition are the two most important things to be immediately improved in order to prevent more than 50 per cent of the prevailing diseases communicable diseases and nutritional deficiency diseases, and to prevent appreciably the morbidity and mortality in infancy and childhood.

Accessibility of health services for curative and preventive care is another thing requiring attention and the government has been doing its best in this regard by expanding the infrastructure. Enough awareness in the community has yet to be created to utilise the services.

Above all education happens to be the most vital area of development. It is through formal as well as informal education that society can be made to understand how to live a better life and avail of appropriate technology for personal hygiene, environmental sanitation, health and family planning. As was pointed out by Dr. Karan Singh in his Sardar Vallabbfiai Patel Endowment Oration recently, primary education for all children is a must, and there should be a sizeable component of health education throughout school and collegue. He also stressed the need for vocational education at the middle level to make education applicable for daily life and occupation and rationalisation of higher education again to make it relevant and useful for society. Needless to say special attention is required for women's development through informal education in the villages.

With regard to the Family Planning Programmes, it may be said that the present strategy for implementation has to be drastically modified. There hits been undue stress on sterilisation at all levels of policy, planning, implementation and review. For having a meaningful change in the demographic picture, the shift in the emphasis in the services should be on providing spacing methods for the young couples. The target-oriented approach with campaigns and camps has resulted in a mad competition among the peripheral workers in health as well as other departments for catching cases for sterilisation.

The programme, if it has to be enduring and have clear impact on the birth rate, should evolve by systematic education. Studies have shown that the health staff themselves are lacking in knowledge about the different contracepthy methods and their attitude towards spacing methods is not yet positive with the result they have a tendency to adjust their records without actually distributing condoms or pills or introducing the IUD, The health staff require to be trained repeatedly by continuing education on two things—contraceptive methods and methods of communication and educations.

It will be better for the government to put a stop to the incentives and compensation money and divert this huge expenditure for improving service facilities. Regarding incentives to staff the question is why pay them extra when they are already paid salaries for doing their normal duties.

Regarding compensation money to the clients the idea was to compensate the loss of wages. This consideration needs to be shown only to the poorer section of the society. There was no need to have extended this facility to government employees and those in the middle and upper classes who are granted leave of absence from work without having to lose their salary. On the whole the strategy should be based on proper education and counselling and with equal emphasis on all methods of contraception.

The educational component should cover more the health aspects because it is very difficult for the couples in the poor sections of the community to perceive or experience the economic aspects. This has been well brought out in a recent study of the ICMR done on a pilot basis in four centres to find out the positive and negative effects of family planning. The small families could not appreciate any improvement in their economic status whereas an improvement in the health of the children could be appreciated.

4 CROWDING

Crowding deals with a subjective psychological state when there are too many people in too little a space. The feeling of crowding is negative and has elements of stress. Typically one or many of the following characteristics are associated with crowding. Feeling of discomfort, loss of control over one's environment, lack of privacy, lack of initiative in soical interaction, fatigue, boredom etc. Crowding therefore, has perceptual affective and behavioural consequences. The outcome of such a phychological state is use of strategies that can reduce the feelings of crowding.

Some recent studies of animal populations in which overcrowding is related to various behavioural pathologies are most suggestive of the stress produced by population density. Calhoun was one of the first to note that disturbances in the territorial and ecological distribution of animals may result in behaviour pathology and physiological malfunctioning. In mice, overcrowding in laboratory cages leads to abnormal sexual behaviour, decreased reproductive and nursing capacities, aborted pregnancies, deficient maternal care for the young and disrupted nest building.

Calhoun reared a colony of Norway rats in four different pens, the rats tended to live together in one very-crowded pen. Calhoun coined the phrase "pathological togetherness" to describe this behaviour which lowered the rats' fertility and shortened their lives. Other biologists have observed that mice crowded together in cages display a marked enlargement of the adrenals, a sign of stress; and there have been reports of a high incidence of cannibalism and male impotence in crowded mice colonies, adjustive mechanisms which function to keep down population numbers when they rise above a critical point. It is not possible to dismiss these pathological reactions as unusual phenomena associated with artificial environmental conditions set up in a laboratory, for such behaviour occurs also in natural environments.

Deevey has reported on the behaviour of lemmings, rat-like animals found in great aboudance in Norway. Approximately every four years, when the population becomes excessive for the available territory, there is a mass migration in which huge numbers race blindly to the sea and drown. Such panic-like reactions to overcrowding in laboratory and natural environments are evidence of regulatory mechanisms that enable population control and, thereby, survival of the species. Whether extreme overcrowding in human populations produces analogous reactions of stress and social pathology is a moot-point. Several models of crowding and its consequences on human behaviour have been proposed and/or studied.

According to overload model, high population density is a stressor as it stimulates the individual, excessively. Too many people in too little space not only crowd the environment about engage in more activities creating more noise, more pollution, more distraction etc. all of which overloads the system. This results in the inability to handle all the incoming information. A consequence of this condition is an aversive feeling. To reduce the overload of stimulation caused by crowding, individuals engage in behaviors that reduce the unavoidable interference from others. At the physical level, the overload leads to fatigue and this is reduced behaviourably by withdrawing.

At the cognitive level, since many cognitive stimuli, have to be handled, it results in confusion. Some investigators also regard excessively stimulating environments as causing stress by reducing the attention span, resulting in impaired cognitive functioning. Less friendly and less helpful behaviors seen among city dwellers is an example of such strategy used for coping with stimulus overload. In homes there is greater dissatisfaction among members and desire to spend time outside is more.

High density conditions create stress by imposing restrictions on behavioural freedom. This limitation can produce a psychological reaction whereby the individual is motivated to remove the restriction by either withdrawing or by compensating for the density by engaging in behaviours that help gain control over the environment. For example in a crowded home or hostel people delineate area or seats for themselves as a means of gaining some control over one's crowded environment.

High density conditions lead to disruptions in accurately understanding the functions of objects in terms of their affordance. All objects in the environment, convey affordances about their functions. Eg. A fruit says eat me; a ball says kick me; a baby says protect me etc. These affordances act as cues for adapting to the environment, efficiently. High density environments block these affordances resulting in the misreading of the cues regarding their functions. For example, a family which should be perceived as supportive and emotionally positive, due to population density is seen more as a nuisance, rather than a source of joy or support. In such a perceived situation the absence of close-warm feelings among members and the desire to form smaller cliques etc. are easily understandable.

Many extended families are now breaking up into separate entities or becoming fully nuclear one of the reasons for this being crowding. The situation is further compounded by the scarcity of social roles and physical resources. For eg. The grand father plays the senior role member, even though the adult sons have taken over the functions of providers for the family. The thwarting of such roles and privileges can lead to a feeling of lack of control over one's environment. Behavioural consequences of this would be to opt for situations where control over ones environments possible, failing which, the stress is experienced without relief. This can give rise to negative feelings about the family.

In attribution model individuals who experience lack of personal space due to crowding and therefore feel the negative attributes the stress to space encroachments. Eg. When too many guests arrive and space is short, there is sense of relief when the people leave. However, there are certain situations when high density is not very negatively perceived Eg. An enjoyable and voluntary party. So the critical element in response to limitations of space lies in attributing the negative feelings to crowding. If the attribution is absent; then the sense of crowding is not experienced, so negatively, even though the environment is a high density one.

How do high density conditions produce stress? Any physical, social or emotional stimuli that are unsettling can be described as stressful. Stressful conditions are characterized by:-

a) Stimuli that are very intense.

b) Beyond the limits of one's control.

c) Threatening, due to one's inability to cope with it.

High density environments span across these three factors making a stressful situation.

Sources of stress have also been identified:

1. *Physical*: Any intense stimulus arouses stressful psycho-physiological responses. Extreme noises, temperative pain etc. are conditions that cause stress.

2. *Psychological*: Deficiencies in personality make up stress the individuality impeding the progress towards the goal. Eg. Inadequate skills, attitudes, poor leadership qualities, having an avoidant personality etc.

3. *Social/Environmental*: Social factors like collective norms traditions etc. can stress the individual by acting as barriers to goal attainment. Environmental factors that stress the individual include, noise, pollution, poor services, over crowding etc. Stress involves s situation where the individual experiences threat to his important goals and also feels that he is unable to cope with the threat. There is also an impending sense of going to be overwhelmed.

Stresses can arise from major traumatic life events like death of a spouse, parent, personal injury, retirement, pregnancy etc., or from the frequent minor hassles of everyday living. Fortunately traumatic events though extremely stressful, are rare. The minor daily stresses make up for their low intensity by their nagging frequency.

Stress affects the physical and psychological health of the individual. It has implications for performance in tasks as well as on social behavior. The effects of stress on physical health has been well researched. Stress is thought to be silent killer. Medical reports suggest that stress affects health to the tune of 50-70%. Stress has been implicated in the occurrence of diseases like heart attacks, hypertension, hardening of the arteries, ulcers and even diabetes. This is because stress renders the immunological system inefficient.

The more prolonged the stress is, the greater is the threat to the immune system of the body. Studies done on individuals who are in mourning or about to take an important examination have been found to show disruptions in their immune system. However, what is encouraging is, that once these high stress levels are lessened, the restoration of the immune system, is rapid. Stress inevitably arouses the organism. This is a survival oriented reaction of the organism to meet the emergency situation induced by the stress. The body's resources are then mobilized to meet the threat. This is a demanding and exhausting process. Therefore, if stress is constant or/frequent it takes a toll on the physical health and psychological well being of the individual.

Crowding qualifies as a constant stressor. In order to escape the harmful effects, the individual engages in behaviours that seek to lessen the density and thereby the experience of crowding. This in turn would lead to a lowering of the stress. Typically this manifests in the form of less helpful and less friendly behaviours. This can also explain the urban/rural divide in terms of friendliness and feelings for others.

The long term effects of crowding have shown higher arousal of the system. Arousal is a form of energization of the system. The greater the stress, the greater the arousal. However, adaptations to the stress also occur. But the adaptation generally involves raising the base line. This means that the threshold for stress tolerance becomes lower. Subsequently in real terms it means that events produce less stress are also, now capable of effecting the organism.

Since the effects of high population density are generally negative, it has consequences for social behaviour. In a study, people who were already in a crowded room were informed that more people are going to come in. The anticipation of more crowding led to feelings of discomfort. It showed up in reduced liking of even fellows waiting alongside. This is seen in a train compartment. Even as the train enters a station the possibility of more people entering produce stress. This leads to overall reduction in friendliness. Significant reduction in helpful behaviour like making room for luggage or staying clear of the passage etc. can be explained in terms of the effects of the crowding.

Individuals who are required to be in environments that are both confining and at the same time be constantly exposed to other individuals as in prisons and other settings like remand homes, free hostels etc. showed greater irritability and reduced satisfaction and harmony in social relationships. In the community settlement the "Kiffutz" in Israel, where children are raised in a community with multiple mothering lacked privacy. This absence of privacy and constant exposure to others resulted in feelings of tension. Anxiety and aggression were some of the other characteristics that were observed.

Students living in three and four seater rooms expressed greater dissatisfaction with their roommates as compared to those living in two seater rooms. The latter also felt less control over their interactions. Some studies have shown that the grades obtained by the students in the more crowded conditions were lower than those from the more comfortable settings. These differences can be attributed to social density. Behaviours that follow, typically involve attempts to reduce the stress and unpleasantness brought about by crowding. Withdrawing and/or less friendly/helpful orientations are the most common forms of response. Table below shows the satisfactions derived from roommates in two conditions.

Satisfaction with Roommate under crowded and uncrowded conditions

	Uncrowded	Crowded
Satisfaction with roommate	4.9	3.7
Perceived cooperativeness of roommate	4.7	3.9

* Higher numbers indicate more positive responses.
(Based on data from Baron et.al., 1976)

The lowered frequency of interaction in high density places can be seen as the effects of less attraction for neighbours. This can then be attributed to crowding whereby lessening of contacts is a way of ensuring reducing the unpleasantness of crowding. Although it appears that high density leads to lower attraction, there is evidence that under conditions of high spatial density, this response is more characteristic of males than of females.

In an experiment male and female subjects participated in same-sex groups of six. Consistent findings on a variety of measures indicated that while males responded more negatively to group members in high than in low spatial density conditions, females, liked group members more under high density conditions. They state that while both males and females experience arousal from high density, social norms permit females to share their distress with others in their group, which leads to liking and cohesion. The same norms prohibit males from sharing stress, which causes a more negative response. In a follow-up experimental, 1976, it was found that when females were not permitted to interact with each other, their positive reactions to high density were limited.

Prosocial Behaviour. If high density leads to lower attraction and to withdrawal responses, how might it affect helping? For example, suppose you lost a contact lens or something else of value. Where would you be most confident of finding someone who would help you look for it; in a high-rise or a low-rise building? In a large city· or in a small town? Interestingly, most research on how density affects prosocial behavior has been done in field settings like these. In fact, one set of studies has compared helping in building characterized by high and low density, and a second has compared prosocial acts in large cities and small towns.

In studies that explored helping as a function of building density, it was found consistently that greater density leads to less helping. Studies compared prosocial acts in high, medium, and low density dormitories. Envelopes, which were stamped and addressed, were dropped in the dormitories, and helpfulness was measured by the number that were picked up and placed in the mail. The results showed that 58 percent were mailed in the high density condition, 79 per cent in the medium density condition, and 88 per cent in the low density condition.

In an interesting corroborative study, the effect of social density on compliance with a prosocial request (printed on signs) for cafeteria users to return their trays to designated areas. It was found that fewer users complied during high density periods. The second line of research, which compared urban and rural helping, provides more equivocal evidence concerning a density-prosocial behaviour relationship. In one group of studies it was found that city dwellers were less helpful and less informative to strangers than rural dwellers. Urbanities were less likely to admit a stranger who needed to use a telephone, less likely to assist a caller who had mistakenly called a wrong number, less likely to correct over payments in making change, and less likely to report theft than rural dwellers. A number of interesting field studies have shown that resident of low-rise buildings are more willing to help others than residents of high-rise buildings.

Withdrawal may be associated with high levels of social contact. Studies have found that withdrawal may function as an anticipatory response to high density, as a means of coping with ongoing high density, and as an after effect. For example, the mere expectation of high social density elicits withdrawal responses, including lower levels of eye contact, head movements away from others and maintenance of greater interpersonal distances. Withdrawal also occurs during ongoing high density interactions: Subjects are more willing to discuss intimate topics under low density conditions, and both children and psychiatric patients interact less frequently as room density increase.

Finally, it appears that withdrawal can constitute an after effect of exposure to density. Males were less likely to volunteer for another experimental session after high social density and groups of males preferred larger personal space and recalled fewer names after exposure to high density.

If density can make people less helpful or want to withdraw; then it is possible that density can also lead to hurting behaviours. Many studies have also established greater aggression and hostility from high population areas. Apart from crowding the high density areas like the slums, require that people have to compete for scarce resources also. For eg. space, water etc. Hostilities can break out frequently in these conditions because they have conditions which are highly tension producing. Competition is a major possible source of interpersonal conflict. This conflict can easily become inter-group conflict under conditions that threaten limited vital resources. These would be heightened by feelings of insecurity due to dissatisfaction of needs, resulting from population pressures.

One study is tallisled that moderately high density conditions led to increased aggression in males, while very low and very high density conditions led to decreased aggressiveness. In another study, aggression was found to increase in boys under conditions of high density. These findings can be explained in terms of competition for limited resources which males, more aggressive as a group. Sex differences have been observed with respect to aggression. Same sex groups of eight members were placed in a room that was either large or small. Males rated themselves as more aggressive in the small rooms as compared to females.

Studies of boys and girls in the class room as well as in anticipation of more cordoned, produced more aggression among males. Spatial density, rather than social density seems to have more aggressive outcomes from males.

Mean Scores on the Feeling of Crowding and Social Behaviour for Outside-Inside Density Groups

Groups		*Mean*	*df*	*t*
Outside	High density	61.00	98	49
	Low density	60.00		
Inside	High density	89.00	48	7.49
	Low density	46.70		

**p<.01

Inside and outside conditions of population density and its effects on social behaviour were studied. It was found that density is a necessary but not a sufficient condition for producing feelings of crowding. But high density within the house had significant effects on the feeling of crowding. The following talk gives a comparison of the behaviour feelings of individual living in high low density conditions.

Social Behaviours and the Feelings of Crowding Experienced by the Residents of High and Low Inside Density Conditions

Behaviour/Feeling	*Density*	*Mean*	*t(48)*
Crowding	High	3.55	3.33*
	Low	1.82	
Privacy	High	3.07	
	Low	1.38	3.07*
Physical condition	High	3.22	
	Low	1.44	4.45*
Interaction	High	3.24	
	Low	1.38	3.20*
Neighbours	High	2.65	
	Low	1.69	3.92*
Insecurity	High	2.67	
	Low	1.24	6.5*
Illness	High	3.44	
	Low	1.71	3.39*
Withdrawal	High	3.88	
	Low	2.46	2.43*

Further, crowding affected health, feelings of security, privacy and generally produced the desire to leave. Negative moods and dislike for affiliation were also related to crowding inside the house.

The feelings of crowding in high and low socio-economic studies groups have been compared, and are given below:

The Feeling of Crowding in High and Low
Economic Status Groups

Dimensions of the Feelings of Crowding	Economic Status	mean	t (48)
Comfortable-uncomfortable	High	1.97	
	Low	2.34	2.64
Good-bad	High	1.98	
	Low	2.48	3.57*
High Congestion-low congestion	High	2.02	
	Low	2.14	0.32
Happy-sad	High	2.08	
	Low	2.48	2.85*
Pleasant-unpleasant	High	1.86	
	Low	2.40	4.00*
Positive-negative High	1.86		
	Low	2.36	3.57*

*P< 05
**P< 01

The following table shows the feeling of crowding under high and low density conditions in terms several dimensions.

The feeling of Crowding in High and
Low Density Groups

Dimensions of the Feeling of Crowding	Density	Mean	t (48)
Comfortable-uncomfortable	High	2.32	
	Low	2.00	2.28*
Good-Bad	High	2.40	
	Low	2.04	2.57*
High congestion-low congestion	High	2.00	
	Low	2.14	0.81
Happy-sad	High	2.38	
	Low	2.18	1.42
Pleasant-unpleasant	High	2.28	
	Low	1.98	2.14*
Positive-negative	High	2.25	
	Low	2.10	1.07

*P<05
**P<02

The Jain findings show that high density in itself does not lead to low degree of attraction towards neighbours; but high density conditions can aggravate aversive feelings if space and resources are extremely limited and have to be shared. This finding has implications for architectural and colony features that have to be kept in mind while designing high density housing localities. In the Indian societal conditions where emotional bonds between relatives, neighbours and community have been traditionally emphasised; the western models which predict low inter personal attraction in high density areas, cannot be fully observed.

In fact it is, the high socio-economic class who are also highly competitive and show less interpersonal attraction for neighbours and others, as compared to the lower socio economic group. The typical helping out relatives, neighbours etc. in times of funerals, weddings etc. is more characteristics of the highly populated, lower economic group. The results thus show that economic factors interact with high density and societal values to moderate the effects of high density living conditions on social behaviours.

One of the most critical questions that can be asked about high density is whether it affects task performance. The answer has important implications for the design of all types of living and working spaces (e.g., schools, factories). Most early studies used simple tasks and were consistent in finding no decrements under high social to spatial density. As seem in the table below both high social and high spatial density led to decrements in complex maze task performance, but these decrements were more pronounced under conditions of high social density. In a field setting, also decrements in complex task performance over time in overcrowded dormitory rooms (three persons in a room built for two) were observed.

*Errors in Maze Performance as a Function of Spatial Density and social Density**

Low spatial density	34.20
High spatial density	37.44
Low social density	32.13
High social density	39.50

In another study poorer complex task performance under crowded conditions was noticed, but there was no impairment in simple task performance. Even anticipation of crowding led to poorer complex task performance. The differential effects of high density conditions on single and complex tasks can be explained using the concept of arousal. Arousal when moderate has the properties of being a motivator. But when the organism is over aroused as in the case of overcrowding, the anxiety is experienced.

Anxiety is detrimental to learning & performance. Complex tasks which need greater attention & problem solving capabilities is more adversely affected than simple tasks. So high density is more adverse for complex task performance than simple tasks. When high density conditions prevail & the expectations of task performance are also low; it yields significantly poorer functioning.

In another study interaction among members was found to be an important factor that produces lowering of tasks performance in a high density condition.

Interaction is necessary for density-related task performance deficits to occur. A final and very important question is whether high density can cause *aftereffects,* as well as immediate effects, on performance. As you will recall, noise is a stressor that has been linked to consequences occurring after exposure, and some studies suggest that crowding has lingering effects as well. For example, it was found that subjects exposed to high density later showed less persistence at working on unsolvable puzzles than subjects exposed to low density. In an attempt to determine whether perceived control would lessen affereffects, some subjects had the option of leaving a crowded room to complete the study in a spacious setting (i.e., perceived control). Although no one took advantage of the option, this group showed fewer aftereffects than the group that wasn't offered an opportunity to leave.

The similarity of these findings to those with noise effects further suggests that noise and high density may affect people in similar ways. Performance suffering under high density, can also be explained as stress induced. Stress has been found to systematically lower performance primarily because it distracts the person from concentrating on the task on hand. Lack of concentration in turn is because of the unpleasantness of the experience. This shift in focus from task to the negative emotions reduces efficiency.

Complex tasks stand to be more readily disrupted by stress as these require greater concentration and therefore suffer larger deficits. Decision making under conditions of stress have been described as resulting in bad choices. The strategies utilised in problem-solving under stress conditions have been found to be poor. Many premature decisions are also associated with choices made under stress. Scanning of the possible alternatives for making decisions has been found to be disorganized under conditions of stress.

A tendency to close the issue on hand as quickly as possible is linked to the presence of stress. Here the individual is motivated to overcome the aversiveness of the stress and therefore quickly engages in actions to reduce it. Premature closure without considering all aspects of a given situation results, owing to the stress effects.

Crowding not only induces stress and thereby lowers performance but can also be implicated in poorer decisions and premature closures. Crowding also introduces the stimulus overload condition. To reduce the overload, individuals engage in hasty and ill-conceived decisions. Moderate levels of stress which have motivational properties of energizing the organism, has significant positive features, in terms of performance. Medium amounts of stress are considered as arousing the organism to better performance through increased mobilization of energy and better attentional focus. However, when they cross the optimal conditions, then the effects are negative and adversely affect performance. Crowding, due to its excessive stimulating features, qualifies more as a stressor and thereby has deleterious consequences on performance.

The only kind of skills that are not very adversely affected are the well established ones like driving, swimming, playing an instrument or speaking a language. Crowding as a stressor, has significant effects on the health of the individual. This in turn has negative consequences for learning & performance. Since the effects of crowding on health is also infavourable, performance can also suffer on account of reduced health in an indirect manner.

The effect of stress on learning is even more disruptive. Learning involves the acquisition of a new set of responses. This implies that the correct response exists at the 0% level of probability of occuring. This low probability response is at the bottom of the hierarchy. Learning means pushing this 0% probability response to one of 100% probability. Such an endeavour requires high degrees of attention and cognitive imputs. Stress induced by crowding, arouses the organism and such an arousal state is energizing.

When learning is engaged in this energized condition, the first natural response to occur is a previously learnt 100% probability one. The new response (to be acquired) has to now compete with the well established earlier response. By the very nature of the process, the newer response has a disadvantage and would therefore suffer, seriously. Research evidence shows that even the mere presence of others is deleterious to learning.

Crowding (which is the presence of many people) would lower learning not only because of the presence of others, but more so because of too many people. Thus, stress induced by crowding would hamper learning in an even more pronounced manner. The new responses to be learnt would in the competition with the older established ones, get depressed easily. So greater effort and attention would be required to offset this condition. Stress in such a context is clearly debilitating.

Education imparted in high density classrooms is found to be ineffective. Further, children who study in classes with large numbers, are also more likely to come from homes that are crowded, than otherwise. This would suggest that both the conditions are unfavourable for promoting learning or that extra efforts would have to be made for reasonable levels of learning to occur. In such strained conditions, students would also not readily have role models of sustained learners. To compound it, other stressors would be in operation in such disadvantaged environments. All these together pose a set of factors that are not conducive to serious learning activities. This could well be the explanation for the absence of children from the slum sections benefitting from long-term education.

Crowding inevitably leads to greater competition for both space as well as resources. If competition is continuous, it can produce increased tolerance for competition according to Uday Jain, who has done pioneering work on crowding in India. Scarcity of resources typically manifests in struggle for power with a view to cornering the scant resources.

In India, the immense desire for power is not so much for power itself but more because power symbolises access to resources. Obtaining railway tickets, getting admissions in educational institutions, jobs, telephone connections, licences for business and industry, etc., are all related to power as a means of need-gratification. In Jain's study it was clearly demonstrated that high density situations lead to greater competitiveness as a way to obtaining the limited available resources. Urban people from high density areas showed more competition tolerance than those from the low density areas.

Similarly, the high density urban individuals competed more than the high density rural individuals. High and low density areas in the rural setting also yielded similar results. In this situation of competing for scarce resources, people hardly find time to think or be mindful of others' needs. This can account for the apparent indifference of the present day city dweller. Given that one has limited time/energy, etc. for obtaining some benefits for oneself, expecting one to be very caring and considerate to others seems to be a very unreasonable one. The blanket statement made about the younger generation perhaps needs to be reinterpreted using the concepts of high levels of competition generated by high levels of population. It is a clear that the earlier generation competed less for the resources as there was a larger amount that could be shared.

If competition becomes fierce as the population increases, it is possibly that the individual's interpersonal relationships would be markedly affected. The biggest causality of such a condition would be consideration for others. Studies show that there are urban/rural differences in the extent to which they tolerate competition. Greater tolerance for competition has also been associated with the need patterns of the high and the low competitive individuals. Low competitiveness is associated with helpful and considerate behaviours while high competitiveness is related to order, neatness, organization, etc. The latters needs are likely to be more readily disrupted in a crowded environment.

The reduced consideration and helping behavior found among the high competitive individuals can be easily attributed to their negative perceptions of others who are viewed as rivals. This is due to the fact that too many persons would not only increase the competition but also intrude into their neat and orderly styles of functioning. Such a situation would naturally yield less consideration and helpful behaviour.

High density dwellers have to encounter vast numbers in their daily activities. They try to conserve their psychic energy by keeping interaction to the barest possible number of people. In such a context, friendly, helpful and considerate social interactions are highly unlikely. This is more clear among the more educated who are more competitive and thus uninterested in nurturing supportive behaviours with others with any degree of spontaneity.

So, the explorations of several density-behaviour relationships shows that high density has various kinds of negative effects. High density is like any other potential stressor and it has !

(1) Immediate effects, such as physiological arousal and negative states.

(2) Coping responses like withdrawal, aggression etc.

(3) After-effects and cumulative effects like health and performance.

(4) Motivational consequences (increased tolerance for competition)

Summary of Theoretical Perspectives on Crowding

Conceptual Approach	*Critical Causes(s) Crowding*	*of Primary Coping Mechanisms*
Social overload	Excessive social contact too much social stimulation	Eacape stimulation; prioritize input and disregard low priorities withdrawal
Behaviour constraint	Reduced behavioural freedom	Aggressive behaviour leave sitution coordinate actions with other
Unwanted interaction	Excessive unregulable or unwanted contact with others	Witdrawal; organization of small primary groups
Interference	Disruption or blocking of goal-directed behaviour	Create structure aggression; escape
Ecological	Scarcity of resources of outsiders	Defense of group boundaries; exclusion

All these models are overlapping and are supplementary in terms of explaining the stress experienced crowding and the behavioural consequences that follow, being stress-alleviating in nature.

Recent conceptualisations on the effects of high density revolve around two concepts control and privacy. Perceived control is regarded as a potent mediator of stress. When we believe that we can control a stressor or other aspects of a situation, the aversiveness of stress appears to be reduced. It follows that even if no other problems are apparent, losing or not having control can also be stressful.

Several researchers have propsed that high density can cause a loss of control (or prevent some one from ever having control) and that this loss of control is the primary mechanism by which density causes stress. Can we really explain the negative effects of high density as a loss of control? The answer is "Yes," even though at times the explanation seems a little strained. Some explanations fit well the *notion of unwanted interaction* is a control-based perspective. Negative effects are the result of contact that is too frequent which makes control over when, where, and with whom people interact difficult to manintain. As a result, interactions become unpredictable and frequently unwanted.

The *behaviour constraint* notion is also control-based, decribing high density as eliminating behavioural options and reducing freedom to behave as one might like. For example, having inadequate space can constrain our behaviour by making it impossible to control the nature of interaction with others. We can generally control the degree of intimacy in one-on-one interactions by adjusting the distance we stand from people but in a very crowded room we may find ourselves with no choice-we must stand close to people whether we know them well or not. In this manner, we can lose control over intimacy regulation.

Interference can also be viewed as a threat to control as our attempts to achieve one or another goal are repeatedly blocked or disrupted. And resource problems the focus of ecological models, can limit our choices and restrict our ability to exercise control. To some extent, all of these consequences of high density that have been related to negative effects do cause a reduction in control. But, is there any direct research evidence for the relation between density and control? Research examining the links between high density and loss of control has taken two different tracks. The first has been to manipulate personal control (provide some subjects and not others with perceived cantrol) and see if it has any effect on experience in high density settings.

Example subjects response to riding in an elevator crowded with confederate of the experimenter. They observed a tendency for people to gravitate toward the elevator floor selection panel (a control panel, if you will), so they manipulated whether subjects were able to stand near it (high control condition) or not (low control condition). The results indicated that subjects standing near the panel felt better than those in a defferent corner. They also thought that the elevator was larger than subjects not near the panel.

A some what different approach found that noise was associated with negative after effects which was reduced when subjects had some degree of control over the noise. A similar, study with density instead of noise was carried out. Once again, negative after effects were associated with stress when control was not available. Subjects who had more perceived control, on the other hand, did not exhibit negative after effects following exposure to high density.

In a high density condition in the laboratory, subjects had varying degrees of control over the setting. Those with control felt better than those without. These studies demonstrate that control can reduce the effects of high density relative to no-control, high density experiences. They do not, however, demonstrate that high density itself has consequences similar to those associated with loss or lack of control. However studies conducted in field settings found that chronically high residential density was associated with helplessness-like behaviour.

Learned helplessness is a syndrome in which people who are exposed to uncontrollable setting learn that they cannot control the setting, and hence stop trying to do so. This manifests itself in reduced motivtion and congnitive activity. Subjects, who were children and adolescents showed symptoms of helplessness that were associated with density in their homes. Symptoms of helplessness were also found among people exposed to high density in their residential environment over prolonged periods. More recently this has been directly linked to perceptions of control in residential settings.

As people relinquished their beliefs that they could control these settings, their behaviour became increasingly like that asssociated with helplessness. All of these studies taken together suggest that control is involved in negative high density effects. When control is available, high density has less impact on pepole than when it is not available. Further chronic exposure to high density appears to be asociated with learned helplessness.

Imagine yourself growing up in a small apartment with five other people who are continuously interacting with each other and with you. With so many people in so little space, you may grow up to feel the world is a complex place in which you have little power to influence events (What consequences does this have? When we come to believe we cannot control our outcomes by responding appropriately (as may result from living in high density conditions), we no longer perform effectively in a number of situations. This syndrome is called "learned helplessness."

Two interesting and provocative studies demonstrated the relationship between residential density and susceptibility to "learned helplessness". In one experiment it was found that children who lived in high density conditions were less likely than those in low density to try to control the administration of rewards they were to receive. In a second experiment subjects from both groups to an initial frustration task on which responses and outcomes were noncontingent children from high density homes did significantly worse on a subsequent task for which outcomes were contingent.

Thus it appears that home density is an important determinant of both the use of control and performance after frustrating noncontingent reward situations. Besides the home environment what other settings may be source of 'learned helplessness' training? One important culprit may be the schools. As classroom size increases, learned helplessness training begins to occur. They argue that larger classroom sizes lead to lower student expectations for control of reinforcement, because teacher feedback concerning student work becomes less descriminative. For example, as class size goes up, individualized student-teacher interactions decrease, and generalized (rather than individualized) praise and criticism increase. Clearly such conditions could lead to a state of learned helplessness with its negative consequences for performance. This could have profound consequences for learning.

An alternative to the density-intensity model has been proposed by Freedman. He agrues that density only intensifies reactions that would occur in any case in a particular situation. High density according to him, heightens the importance of other people and magnifies our reactions to them. Thus high density would intensify the aversiveness of an unpleasant situation and intensify the pleasantness of positive situations. From this point of view, a number of factors can lead to a negative reaction in a high density situation.

One of the benefits of such a model which utilises the notion of intensfication; is the link between crowding and contagion. Contagion is the rapid spread of behaviours or emotions from one person to the group. Eg. People coming back after viewing a game. The negative effect, (if the home team loses) would be quickly passed on among all those returning. This model provides a frame work that specifies three factors as important in determining the experience of crowding. They are:

(1) Individual difference

(2) Situational conditions

(3) Social Conditions

Identifying individual difference variables can help determine who among the selected persons will be more sensitive to the constraint of space. Studies have already established gender differences in the aversire feelings, due to crowding. Some persons prefer to sit for away from others while others prefer to sit closer. This can also affect the degree to which crowding is experienced.

Personality characteristics also moderate our reactions to high density. It has been found that internals (who feel they control their fate) display a higher threshold of crowding than externals (who feel events are controlled by outside forces). The primary problems encountered in high density settings may also affect the ways in which individual differences are important. People who screen themselves from interaction and organize their surroundings were better able to cope with social density than were people who did not screen themselves. One would expect that this variable would be less important when spatial problems were paramount.

A final individual difference variable that has been linked to reactions to high density is one's adaptation level (from past experience under high density conditions). A number of investigators have hypothesized that people with a history of high density living are less likely to experience crowding in a novel situation than are people with a history of isolation. For example it has been found that the Japanese, residents of HongKong and the Logoli (all of whom live under extremely high density) have developed social mechanisms that may be viewed as adaptive for high density living. In addition to being moderated by individual difference variables, our reactions to high density are also affected by *situational conditions*.

An important situational condition is the degree of control we have. Complementing those discussed earlier, several additional studies have found that allowing people increased control over a situation, leads to less perceived crowding. The applied potential for introducing control in high density conditions, is phenomenal. For e.g. Individuals living in high denity situations, can be given training for enhancing control. ie. how to share space and ensure privacy. This is bound to help alleviate crowding and it negative effects.

Other situational factors as whether we are in a primary environment (like a home) or a secondary environment (like a restaurant), and the extent to which other stressors, such as noise etc. are involved, also affect crowding. When we are in a primary environment we may be more likely to experience crowding. Finally, in terms of activities there is suggestive evidence that we are more likely to experience crowding when engaged in work, rather than in recreation. Life styles that increase social contacts like picnics, games, travel etc are more likely to generate feelings of crowding in dense environments than life styles that decrease the likelihood of social contacts.

In addition to individual differences and situational conitions, *social conditions* can be manipiulated to affect whether or not we are crowded. These are variables that make up the social "climate" of a high density situation such as the degree of friendship and the level of social interference and interaction. For example our relationship with the people were with, may determine how crowded we feel: Less crowding is experienced with liked rather than disliked others and with acquaintances than with strangers.

To the extent that we experience social interference (interruptions) by others or excessive proximity or intimacy (too direct eye contact or body orientation) crowding is more likely. Finally crowding is more often experienced in unstructured task situations than in structured ones. High density within primary groups (e.g. families) has repeatedly shown fewer efects than it has in other groups—when density *within* apartment units in examined, for example it appears to be negligible as a factor associated with illness or behavioral difficulties.

Degree of acquaintance and relative position of an individual in a group's dominance hierarchy also affects crowding. The presence of friends or the possession of high status tends to blunt the aversiveness of large numbers of people or cramped spaces. Since some of the conditions and factors that mediate the experience of crowding have been delineated; it is worth considering how one can modify existing environments or plan new ones to mitigate the effects of crowding.

To begin with, greater ceiling height is asssociated with a feeling of less crowding. Rectangular rooms seems to elicit less crowding than square rooms, of the same area. Rooms that have visual escapes such as doors or windows are rated less crowded than rooms that do not have such escapes. The adding of flexible partitions in rooms, lessens feelings of crowding. Increasing the brightness of rooms by providing suitable wall colors leads to less perceived crowding. The presence of visual distractors e.g. pictures, some movement etc. also helps reduce crowding. Finally, studies that have evaluated the aversive effects of Social *vs* Spatial density have established that social density is more negative than spatial density. Architects and designers must address themselves to the fact that too many people produces greater sense of crowding than too little space.

Cities are the hotled of crimes of different kinds owing to a peculiar mix of people, social, physical and housing factors. Large scale rural to urban migration; the concentration of the underprivileged in core unban areas, crowding of population into higher and higher densities all significantly affect life styles environmental conditions and crime rates, in the large metropolitan areas. It is now becoming increasingly clear that the physical and architectural designs of residential environments can elicit attitudes and behaviour from residents which contribute in a major way towards insuring their security Residential environments with certain physical features have crime-inhibiting qualities.

As a starter, 'semi-private' spaces should be 'designed in' to facilitate social interaction between residents and thus encourage the development of social order. Such areas should exist both inside and outside the building and should be planned so as to make pleasant social interactions inevitable. This is specially true of many South Asian countries and particularly India, where dependence and interdependence rather than independence is a valued social goal in interpersonal relationships. Many homes have extended families, which involve the living together of many individuals of different generations. So our housing colonies should be totally designed to suit our socio-cultural environmental therefore needs. So houses that alienate cannot be appropriate for India. Yet, colonies that grow up where spaces are 'nobodys' can become dumping grounds easy for encroachments or neglect and subtly support crime and randalism. In order to take care of this situation the design of a 'defensible space' has been proposed. This notion implies that building should be so designed that all the space appears to 'belong' to some individual or group. Such areas can be easily depended and therefore promote surveillance and thus reduce unauthorised activities and crime.

In practice this means that areas around buildings e.g. service lanes, parks playgrounds etc. must belong to the building. Then these areas would be perceived by both the residents of the building and outsiders as extensions of the building. The otherside of this issue is that public areas such as hallways, corridors, stairways etc. should be eliminated as far as possible. This can be done through horizontal rather than vertical constructions. Where this is not possible then open 'air spaces' like courtyards combined porches or hallways that are accessible to only a small group of residents should be constructed. These designs would facilitate territorial functioning. If corridors have to be provided; they can be made to run along the window side. This is easy to be monitored.

5 Population Characteristics

The population growth in developed regions during the 18th and 19th centuries was due to decrease in mortality rates following substantial improvements in life-style, living standard and environmental sanitation. Rise in the population growth and consequent rise in per capita product, however raised some important issues. The relationship between population growth and socio-economic development is very complex. It is difficult to make a generalization on this matter, because the experiences of different countries vary a great deal. As far as policy implications for developing countries in the light of the experience of developed countries are concerned, it may be stated that a high population growth rate is not desirable for social and economic development.

A vast majority of the population of developing countries live in rural areas in a state of poor economic conditions arising mostly from increasing landlessness and falling real income. There are various direct and indirect ways of creating positive relationship between poverty and fertility. For example, it has been well established that mortality, particularly infant/child mortality is inversely related to the socio-economic status. Consequently, a high mortality in many developing countries has been considered one of the great hindrances to the adoption of small family norms for the reason that high fertility is required to offset high mortality. It is, therefore, hypothesized that to reduce fertility and mortality there is a need to improve the economic conditions of the rural masses in developing countries. The consequences of high population growth on various aspects of social and economic development is discussed below with a special reference to the Indian situation. India's population problem may be shown through (a) its absolute size (b) rate of growth of the population, and (c) its age structure.

1. According to 1991 census, the total count of the population of India was about 844 million. The average annual growth rate during the decade 1981-91 is 2.14% as against 2.20% and 2.22% respectively during 1961-71 and 1971-81. The large sized population itself accentuates the problems of social and economic development by the demand it makes on consumer goods and capital.

2. Because of this broad base, even a low rate of population growth adds a large population to the country each year. The geometric rate of population growth in India during 1981-91 was 2.1 per year. Although this growth rate appears to be low, it has a tremendous effect on population increase. Hence from the point of view of population growth, the growth rate of 21.1 per year may be classified as contributing to rapid population growth.

   Thus a large size of the population and its rapid growth are the most important characteristics of India's population and they have significant impacts on the country's social and economic developments.

3. However, population does not become a problem for the nation because of its large in every case. The age structure of the population is also an important aspect of the problem. For example, if a country's

birth rate is high, then each year a large number of non-productive persons will be added to the population. Thus, if the population base is large, as is the case in India, the yearly number of new entrants is pretty large. Further a declining infant mortality rate in many developing countries increase the number of children in the total population. Nations with high rates have 14% to 20% of population which is under five and 40% to 50% under 15 years of age. This implies that the dependency ratios in developing countries are higher than that in developed ones.

After the second world war, several Asian African and Latin American countries became free from colonial rule. With the achievement of political independence came heightened hopes and aspirations for eradicating poverty and wiping out the evils and miseries associated with poverty. These independent nations also aspired to raise the standard of living of the masses in order to improve the quality of life. They chose the path of planned development and adapted time-bound national development programmes for the achievement of social and economic betterment of their people.

However a high rate of population growth nullified the efforts to improve the quality of life. Boserup hypothesized that population growth rather than being a hindrance to economic growth is actually a pre-requisite for development as it also modifies the population-resource ratio through out-migration. Needless to mention that the world being global village, it was expected that international flow of things like essential commodities, ideas and technologies will alleviate the pressure of wearisome population and remedy the population-resource imbalances evertheless, as mentioned above, the quality of life or standard of living of a country is largely affected, among other things by the population increase in relation to natural resources and the level of technological development. Most of the countries which have a large population and comparatively a higher rate of population growth generally have a lower standard of living and hinders development by raising the depending ratio. That amount of national income which otherwise could be available for augmenting economic and social facilities and raising per capita income for smaller numbers get distributed over large numbers.

Rapid population growth, thus, plays a significant role in shaping the pattern of economic development in the Third World. However, divergent views have been expressed on the precise role of population change in the development process illustrating thereby the point that economic and demographic growth are not exclusive processes. It is, however, important to know whether rapid population growth reduces the pace of development or accelerates the implementation of more efficient development policies.

The problem of interrelationship between population and resource development at any particular point of time owing to the multiple factors involved in it and the role of changing international situations is intricate. Rapid population growth and its effects on social and economic development stimulated demographers, sociologists, economists, development planners and administrators to study in great detail the interrelationship between population growth and socio-economic development. These studies include relationship between national and per capita income, resources and productivity, saving and investments, manpower and nutrition, health and education, social welfare and security status of women, child development and domestic environment, etc. All such studies both at empirical and theoretical levels give a clear view of the relationship between population growth and development. It has been common knowledge, therefore, that a high rate of population growth obstructs development efforts and complicates the existing problems associated with development.

A significant change has been noticed in the Indian agriculture since independence. Foodgrains production has increased from 53 million tonnes to 172 million tonnes during 1988-89. As a result, there has been no need to import foodgrains for the domestic requirement. Although share of agriculture in the Gross Domestic Production has been declining in the recent past, it is still the lead sector in the Indian economy contributing 30 per cent of the GDP, providing employment to 60 per cent of workforce and sustenance to 35 per cent of the population.

All the efforts of self-sufficiency in foodgrains production are being minimised by the rapid rise in the population of the country. According to an estimate, the population of India is expected to reach the level of 94 crores by the year 1997. This will increase to 110 crores by 2007. With this population, the estimated foodgrain requirement for the year 1997 will be around 208 million tonnes and upto 2007 it is expected to be 283 million tonnes. The required development in agriculture, can be achieved only through improved technology of production which is based on the availability of such resources as fertilizers, irrigation, implements, plant protection chemicals, high yielding varieties of seeds and adequate knowledge of cultural practices.

The Government of India has been trying her level best to provide various facilities to the farmers through five year plans and other developmental schemes. The main hurdle in the adoptability of the new technology by the farmers has been traditional cultural practices.

Economic profitability of the new technology did not necessarily motivate many farmers to adopt them. They reject many of these practices which have obvious qualities to be of economic benefit to them. Gave several examples to show how incompatability with local norms, values and habits contributed to the failure of recommended new practices in agriculture to be accepted by the farmers.

The main task before the researchers is to find out how far the new technology of production is being adopted by the farmers. There are various categories of cultivators in India which are based on the size of land holdings. The average size of holding has declined from 3.1 hectares per household in 1953-54 to only 1.7 hectares per household in 1981-82. In addition, it has been found that for India as a whole, the percentage of marginal and small farmers has increased from 60 per cent in 1953-54 to as much as 75.3 per cent 1981-82 The marginal and small farmers because of the limited assets have not been able to get benefit from the new technology to the extent it was expected. Many of them often are not in a position to adopt the kind to technology as would ensure high productivity. As a result, large percentage of them suffer from poverty and ignorance particularly in agriculturally backward regions of the country. Their condition continues to be pitiable inspite of the best possible efforts to raise their standard of living. Even, they could not attract the attention of the scholars and consequently, few studies have been conducted to know the problems of the small and marginal farmers.

The present inquiry is an attempt in this neglected area of research. It proposes to study the extent of adoption of wheat technology by the small farmers at micro level in the light of their population characteristics which directly or indirectly influence the adoptability to a considerable extent.

The present inquiry has a two-fold objective. One to find out the population characteristics of the small farmers, and two, to find out extent of adoption of wheat technology by them. The specific hypotheses that are proposed to be tested in this study are:

(i) Most of the small farmers are either illiterate or have poor level of education, belong to nuclear families rather than joint, come from lower income groups and have sufficient knowledge regarding high yielding varieties of wheat.

(ii) The small farmers have low degree of adoptability in respect of wheat technology.

Naraini block of the Banda district in Uttar Pradesh has been selected for the present study. The block is situated in the southern part of the district. The population of the block is 160788 living in 146 villages as per census of 1991. Most of the area in the block is either unirrigated of partly irrigated by the canals. However, means of minor irrigation have been adopted wherever possible.

Of the 146 villages in the Naraini block four were selected randomly for the intensive study of the problem. These are: Nahri, Piprahari, Palhri and Oraha. After the selection of villages, 25 respondents frorry each village were selected randomly for proper investigation and collection of data. In all, 100 small farmers were interviewed with help of an interview schedule specially constructed for the purpose. The interviews were conducted in January, 1990. Adoption intensity index' was used from the measurement of recommended agricultural technology in the cultivation of, wheat. For this, following 9 practices have been used. The procedure adopted under each practice is given below.

(i) Extent of adoption of H.Y.V. of Wheat:

$$\frac{\text{Area under HYV of wheat (in acres)}}{\text{Total area of wheat (in acres)}}$$

(ii) Extent of adoption of fertiliser

Applied dose of fertiliser (each type) per acre

Recommeded dose of fertiliser (each type) per acre

(iii) Extent of time of top dressing for area

S.No.	Time of top dressing	Scores	Percentage alloted according to score
1.	Correct	3	100,00
2.	Some what correct	2	66.67
3.	Incorrect	1	33.33

(iv) Extent of adoption of irrigation

Frequency of irrigation	Scores	Percentage alloted according to score
1.	1	16.67
2.	2	33.33
3.	3	50.00
4.	4	66.67
5.	5	83.34
6.	6	100.00

(v) Extent of adoption of seed rate:

S.No.	Seed rate dressing	Scores	Percentage alloted according to score
1.	Correct	2	100.00
2.	Incorrect	1	50.00

(vi) Extent of adoption of time of sowing:

S.No.	Time of sowing	Scores	Percentage alloted according to score
1.	Timely sowing	3	100.00
2.	Some what timely	2	66.67
3	Untimely sowing	1	33.33

(vii) Extent of adoption of method of sowing

S.No.	Method of sowing	Scores	Percentage alloted according to score
1.	By broadcasting	1	33.33

2.	Furrow or Pora and traditional method	2	66.67
3.	Seed drill/seed cum fertiliser drill	3	100.00

(viii) Extent of adoption of depth of sowing:

S.No.	Depth of sowing	Scores	Percentage alloted according to score
1.	Incorrect	1	33.33
2.	Some what correct	2	66.67
3.	Correct	3	100.00

(ix) Extent of interculture (Weeding):

S.No.	Method of weeding	Scores	Percentage alloted according to score
1.	Noweeding	0	00.00
2.	By Khurpi/ traditional method	1	50.00
3.	By chemical or How method	2	100.00

Following the above procedure, the intensity of adoption for each farmer was calculated. The minimum adoption intensity of wheat technology in the Naraini Block was 22.27 and 69.42 was the maximum. On the basis of the upper limit of adoption intensity for the small farmers and their range of adoption they have been categorised as given below:

S.No.	Categories	Range of adoption	No. of small farmers
1.	Low adopter	22.22 - 37.95	16
2.	Medium adopter	37.95 - 53.68	43
3.	High adopter	53.68 - 69.42	41

Education is one of the dominant factors affecting each sphere of social life. It makes the people more award of the modem methods of farm production. In fact, it helps a lot in dissemination of the knowledge about modem technology. The small farmers coming in the purview of this study have been classified into 3 categories and the data relating to it has been presented in the following table.

Table 1

Education of small farmers

S.No.	Education	Small farmers (%)
1.	Illiterate	49
2.	Below High School	32
3.	High School and above	19

It is obvious from the table that nearly half of the small farmers are illiterate and about one third of them got education below the level of high school. As against this, only 19 per cent were educated up to high school or above. This indicates that the level of education of the small farmers is sufficiently low.

Land holding is one of the major factors which is responsible for the jointness of the family. The occupation of land cultivation requires more and more hands at the time of sowing and harvesting and thus, for the big landholders, joint family is must. So far as the small farmers are concerned, their holdings are small and, therefore, nuclear Camilics may be common among them. The family types found in the present study have been presented in the following table.

Table 2

Family types of the small farmers

S.No.	Type of Family	Small farmers (%)
1.	Nuclear	73
2.	Joint	27

The table makes it clear that most of the families among the small farmers are nuclear and very few i.e. only 27 per cent are joint. It has already been stated that small holdings require lesser hands and as a result, nuclear type of family has become a characteristic of the small farmers.

Now-a-days, agriculture has become an industry which requires proper investment. It has been observed that the use of improved seeds, fertilizers and irrigation facilities have become prerequisites for higher production, This is possible only when farmers have higher income to invest. The small farmers having smaller holdings are generally expected to have lower income. The income of the small farmers has been presented in the following table.

S.No.	Income (Rs.)	Small farmers (%)
1.	<10,000	53
2.	10,000 - 20,000	30
3.	>20,000	17

It is obvious from the table that more than half of the small farmers have poor level of income who earn annually less than 10 thousand rupees whereas 30 per cent farmers earn in between 10 and 20 thousands. As against this, only 17 per cent farmers were earning more than Rs. 20,000. The small farmers falling in the highest income group were engaged in side occupations besides agriculture. On the whole, most of the small farmers are economically poor.

One of the major factors responsible for the adoptability of new technology is the diffusion of scientific knowledge regarding high productivity. The main task of the extension worker is to diffuse agricultural innovations among the farmers. These include improved implements, high yielding and disease resistant seeds, chemical fertilizers and plant protection chemicals such as pesticides, fungicides and herebicides.

Utility of an innovation refers to the degree to which it is perceived as useful and beneficial by the farmers. The knowledge gained in the laboratory will be of no use if it does not reach to the farm. In the present study, data have been collected regarding the knowledge of the small farmers in respect of high-yielding varieties of wheat and the same are presented in the following table.

S.No.	Knowledge categories	Small farmers (%)
1.	Poor	16
2.	Sufficient	54
3.	Profound	30

The table makes it clear that more than half of the small farmers have-sufficient knowledge regarding high-yielding varieties of wheat and nearly one third farmers of the total sample have profound knowledge. As against this, only 16 per cent of them had poor knowledge. The overall analysis of the data indicates that the small farmers are aware of the new technology in respect of high yielding varieties of wheat.

The load of rapidly increasing population has forced for more and more production of foodgrains which is possible only by adopting new technology.

Table 5

The extent of adoption of wheat technology

S.No.	Categories	Small farmers (%)
1.	Low adopters	16
2	Medium adopters	43
3.	High adopters	41

The table clearly indicates, that the farmers are adopting the wheat technology to a greater extent. 41 per cent of them are high adopters whereas 43 per cent are medium adopters. In contrast, only 16 per cent are low adopters. Surprisingly, not even a single small farmer could be traced with zero adoptability. During investigations and data collection, it has been observed that the small farmers were facing so many difficulties in the adoption of new technology. Poor irrigation facilities, paucity of funds to purchase fertilizers and improved high yielding seeds, adulteration in fertilizers, lack of proper guidance are some of the major hurdles in the way of adoption. In such circumstances, one may easily understand that the adoptability of the small farmers is more than it ought to be. Here, it may be mentioned that our findings find support from the findings of Ratan Chand who finds that about 80 per cent farmers were the adopters of fertilizer in their wheat crop. As against this, they differ from the findings of Chattopadhyaya who concludes that the adoptability of the farmers is low whereas we conclude that more than 80 per cent small farmers are either high or medium adopters and only 16 per cent are low adopters.

Our hypothesis that the small farmers have low degree of adoptability in respect of wheat technology, is rejected. Most of the small farmers are either illiterate or have poor level of education; they mostly belong to nuclear type of families; their income level is poor and they have sufficient as well as profound knowledge regarding high yielding varieties of wheat. The small farmers are mostly high or medium adopters of wheat technology and the ratio of low adopters has been found very low. The non-adoptability is nil. The findings of this study have far reaching implications for the foodgrain policy.

6 Population Policy

It has been appreciated that particularly in India with 15 per cent of the world population, only 2.4 per cent of the world land area, and 1.5 per cent of the world income the high growth-rate of population poses tremendous problems of socio-economic development, not only for the maintenance of present standards of living, but also for raising them. Large increases in agricultural and industrial growth have already been neutralised by rapid population growth. For example, the net national product in 1950-51 was 167,310 million rupees and it rose to 456,370 million rupees in 1978-79. This was 173 per cent improvement. But the per capita income during the same period increased by only 53 per cent, about one-third of the increase in the net national product.

The Indian family planning programme depends for its success on voluntary acceptance by the people, though a process of motivation and education, of the good that accrues to individuals practising the small family norm. The philosophy behind the programme is as follows:

(a) The community must be prepared to feel the need for the services in order that those may be accepted when provided.

(b) Parents alone must decide the number of children they wish to have.

(c) People should be approached through the media they respect and through their recognised and trusted leaders.

(d) The services should be made available to the people as near to their homes as possible.

(e) The services will have greater relevance and effectiveness if they are made an integral part of medical and public health services, especially of the mother and child health programme.

In the early years of the family planning programme, the Government of India had adopted a "clinic approach" to family planning. It opened a number of clinics and hoped that the people would take advantage of the facilities. This approach was modified in 1962-63 to the "extension approach". The Government started giving greater emphasis to motivation and to changing the attitude, norms and values of the people. Social workers and auxiliary nurse midwives (ANMs) were appointed to carry out the motivational work. Other staff, both medical and paramedical, was also appointed; the programme was made target-oriented and additional funds were made available.

Expenditure in 1962-63, for instance, was more than that incurred in the First and Second Plan periods combined. Since the Fifth Plan, the family welfare programme has been integrated with health, maternal and child care and nutritional programmes. This approach is called the "integrated approach".

In India there are roughly 112 million reproductive couples who need family planning advice and service. These 112 million couples are spread out in 576,000 villages and 3,245 towns and cities spread over 1.23 million square miles of area having different terrain, geography, climate and means of communication. Of these nearly 67 million have three or less children and 43 million have four or more children. Family planning knowledge and attitude studies carried out in India have brought out that while very few among those who have three children or less are in favour of family planning, a large majority, between 70 and 80 per cent, of those with four or more children do not wish to have any additional child.

If correct knowledge about family planning is given to those couples and if contraceptive service is made available to them, a large majority of them would take to the regular use of contraceptives. However, besides the problem of physical inaccessibility, widespread illiteracy, specially in the rural areas, lack of adequate media of communication, diversity of language and dialect and general apathy towards any social change among traditional people have made the task still more difficult.

An organisation has been developed to provide contraceptive information and service to the people-the Primary Health Centres-and Sub-Centres as units in rural areas, and the Urban Family Welfare Planning Clinic as the unit in urban areas. A Primary Health Centre covers a population of 80,000 to 100,000 and a Sub-Centre, a population of about 10,000. However, the Planning Commission has approved establishment of more Primary Health. Centres and Sub-Centres to achieve the ultimate norm of one Primary Health Centre for every 50,000 population and one Sub-Centre for every 5,000 population. An Urban Family Planning Clinic covers a population of about 50,000. In the rural areas a rural family planning clinic which forms a part of the Primary Health Centre has medical, paramedical, extension and statistical staff.

The Primary Health Centre has a number of Sub-Centres each being manned by an auxiliary nurse midwife (ANM) and an attendant. In the urban areas, there is a Family Planning Welfare Centre for every 50,000 population. Quite a few of these urban centres are run by local bodies and voluntary organisations. In cities with four or more urban centres, a City Bureau is provided for the supervision of the work of these centres. Steps have been initiated to bring the urban centres and maternity and child health centres together.

At the district level, a District Family Planning Bureau forms a part of the district medical and health organisation and supervises the work of the urban and rural centres in the district. It is headed by a District Family Planning Officer, who is assisted by an extension educator, a mass media officer, medical and paramedical staff for services, a statistical officer and an administrative officer.

An audio-visualpublicity vehicle is also provided to each district, as also a properly equipped mobile IUCD-cum-Sterilisation unit for each 50,000 to 75,000 population. At the State level, a State Bureau is provided as a part of the Directorate of Health Services. The Bureau is beaded by a State Family Planning Officer, who is of the rank of a deputy, joint or additional director. He is assisted by a mass media officer, a health education officer, medical officers for IUCD and other programmes, statistical and evaluation officers, an administrative officer and a complement of secretariat staff.

In order to provide adequate leadership and to bring about necessary co-ordination, a full-fledged Department of Family Planning was created in April 1966 in the Ministry of Health and Family Welfare, Government of India. The Secretary to the Government, Ministry of Health and Family Welfare, is in over all charge of the Department. He is being assisted by an Additional Secretary and Commissioner who heads the family welfare structure, and a Joint Secretary. The Secretariat Wing of the Department looks after the policy, planning, budget, grants, foreign assistance and administration. The Technical Wing looks after training, progress, evaluation, research, mass media, transport and contraceptive services.

Cabinet Committees at the Central and State levels have been constituted to give proper direction and policy clearances and to regularly review the progress of the programme. The Central Cabinet Committee is headed the Prime Minister and the State Cabinet Committees by the Chief Ministers of the States. The Central Family Planning Council is the chief policy-making body and is presided over by the Union Minister for Health and Family Welfare. It includes the Health Ministers of all the States and representatives of voluntary organisations. Similar councils also exist in almost all the States.

The Family Welfare Programme is a centrally sponsored scheme and the State Governments/Union Territories receive 100 per cent assistance from the Central Government. The States, however, are responsible for administering the programme. IUCD and sterilisation services are being offered through mobile and static agencies as well as through camps. Nirodh distribution is organised through three channels. The first scheme provides for free distribution through all the family planning centres and subcentres.

Under the second scheme, called the Depot Holders Scheme, a nominal price of 5 paise is charged for three pieces of condoms and is retained by the depot holders who are honorary workers. The third scheme relates to the distribution of Nirodh through commercial channels. This scheme was launched in October 1968 and presently twelve of India's largest and most experienced private consumer-goods marketing organisations distribute

Nirodh through their network of salesmen, wholesalers and retailers. These condoms are sold at a subsidised retail price of 25 paise for three. Oral pills, one of the conventional contraceptives, were introduced in the Family Welfare Programme in August 1967 as a pilot project. The result was encouraging and the method was included as one of the contraceptives under the National Family Welfare Programme. Oral pills are distributed free of cost. To accelerate the usage apart from registered medical practitioners, para-medical personnel are also allowed to distribute oral pills to acceptors under the guidance of qualified doctors.

Laparoscopic tubectomy, which has been introduced recently, is becoming increasingly popular among women. In order to cope with the demand for trained personnel in the laparoscopic technique of tubectomy, training arrangements have been strengthened. Various States have set up training facilities locally for training their medical and para-medical staff.

In India, the inverted red triangle has been adopted as the symbol of family welfare. This symbol is widely displayed in urban as well as rural areas. All available mass media are being used for disseminating information about family welfare among the masses. The Media Unit of the Ministry of Information and Broadcasting gives considerable support to the Family Welfare Programme through the issuance of press advertisements, production and distribution of a large number of information and educational publications including posters, pamphlets, etc. The Family Welfare Programme is also projected through a number of programmes of All India Radio and Doordarshan.

The AIR broadcasts over 5,000 items every month promoting family welfare. The commercial circuits of Doordarshan are also used for promoting the use of

Nirodh. Some of the publicity units of the Directorate of Publicity are devoted exclusively to family welfare programmes and for the other units, publicity of family welfare programmes is a major activity. Apart from these, the person-to-person approach is also being utilised with emphasis on group meetings. The involvement of the community is sought through the participation of local leaders and voluntary workers.

The family planning programme in India was adopted as an official prow gramme in the early 1950s. It was reorganised in the Third Plan after the publication of the 1961 Census results which showed a higher growth rate than anticipated. The programme was made time-bound and target-oriented. It was only in 1962-63 that the Government specified the objective of the Family Planning Programme to reduce the birth rate from the level of 41 to 25 per 1,000 population by 1973. It is not known how the above figure was arrived at but it appears that if the fourth and higher order births are curtailed, which constitute around 40 per cent of the total births, the above figure can be obtained. This target was fixed at the national level. However, for the States the targets were fixed in 1965-66 and since then the programme has become time-bound and target-oriented.

The family planning target of birth-rate reduction has undergone a number of changes since it was first laid down. For example, in 1966-67, the Government revised the objective of achieving a birth rate of 25 by 1973 and instead stated that it was to be achieved "expeditiously". The objective was changed many times on the basis of performance. The birth rate cannot be brought down by family planning alone. It is also associated with health, social and economic conditions of society. As such, a new exercise was conducted to achieve the ultimate goal of the family welfare programme, that is, *Hum do, Hamare do*

(We two, Ours two). The same in demographic terms can be translated as NRR (net reproduction rate) of unity. This exercise encompassed general health conditions represented by the expectation of life at birth, social conditions through marriage pattern and economic conditions by infant mortality. The details are given in a separate monograph. This exercise was done on behalf of the Working Group on Population Policy constituted by the Planning Commission, Government of India.

The recommendations were discussed in the meeting of the National Development Council and on the basis of the approval of the Council, the long-term demographic goal was incorporated in the Sixth Five Year Plan as the objective of the population policy. According to the Plan document, the above objective of the population policy can be achieved within the stipulated time period "only by reducing the birth rate to 21 per thousand of population, death rate to 9 and increasing the proportion of couples protected by family planning to about 60 per cent".

On the basis of the targets of birth rate reduction, the annual targets in terms of sterilisations, IUD insertions and users of conventional contraceptives were also fixed. Up to March 1981, a total of 33.44 million sterilisations were performed since inception and 8.78 million IUCDs were inserted. In addition, in 1980- 81, there were about 3.79 million users of conventional contraceptives, mostly condoms. Thus, out of an estimated 116.35 million eligible couples in 1980-81, about 28.32 million couples or roughly 24.3 per cent eligible couples were protected.

Depending on the use effectiveness of the method, which is taken as 100 per cent for sterilisation and oral pills, 95 per cent for IUD and 50 per cent for conventional contraceptives, the number of couples effectively protected works out to 26.41 million or 22.7 per cent of the eligible couples. A result of the family planning efforts in India, there has been a marked decline in the level of birth rate. Prior to 1965, there was little work done in the country in the field of family planning. Since then, during 1965-71, about 6.45 million births are estimated to have been averted accounting for a 2 point decline in the birth rate during the above period. If the birth rate during 1961-71, centred around 1966, was 41 per thousand population, then it should be around 39 in 1971. Again, on the basis of the provisional population totals of the 1981 Census, an exercise was made which shows that between the years 1971 and 1981, the birth rate had declined by about 4 points. Thus, in 1981 the birth rate in India was around 35 and the decadal rate for 1971-81 works out to around 37. A major portion of this decline is attributed to family planning efforts. Budget provision and expenditure on family planning have increased considerably since the beginning of the Third Five Year Plan, when the programme was reorganised. While the budget provision in the first two Five Year Plans was less than 500 lakh rupees, it was increased to about 2,697 lakhs in the Third Plan.

The budget provision in the subsequent Plans was further increased. In the Sixth Five Year Plan, provision for 101 thousand lakh rupees has been made for the family welfare programme. Like the budget provision, expenditure on the family welfare programme has also increased considerably. So far the family welfare programme in India has achieved only limited success. While it is true that a large proportion of the population is in favour of family planning and that about one-fourth of the eligible couples are currently protected by various methods of family planning, much more serious efforts will have to be made if the birth rate is to be reduced to the level of 21 per thousand population by the year 1996.

To achieve this target about 60 per cent of the eligible couples will have to be effectively protected against conception—and this is not an easy task.

The complexities involved in the solution of the problems of rapid population growth in a developing country like India necessitate originality and initiative in evolving new approaches. The past trends, which are well-documented for the more developed countries, will not provide the necessary means for solving the population problems in most of the developing countries. For example, past trends do not provide us with even the basis for perfectly understanding the social and economic factors affecting fertility beyond the direct effect of increasing per capita product and associated changes in the economic structure. While present population trends will not be a mere repetition of the past, the consensus among experts in the field indicates that the theory of demographic transition is still valid.

There are different opinions on how any Government should advance through the various stages. Some think that sufficiently extensive national and international efforts aimed at improving social and economic conditions would reduce the birth rate and population growth without direct Government intervention in the field of family limitation. On the other hand, others are of the opinion that there is a need or desirability of systematic Government action in encouraging; the adoption of small family norms to lower the birth rate. Such differences in approach are likely to continue, and it is only through experience that each Government will know which of the policies is more rewarding.

Indian planners think that family planning involves a change in the norms, and in the attitudes and values of the people so that they come to prefer small families in place of the traditional large ones. But, unless there is a change in the main aspect of the traditional way of life, it will be difficult for people to change their attitudes and start favouring small families.

Family planning, therefore, must precipitate social change. In the countries of the West, where governments did not find it necessary to promote and finance family planning, the decision to restrict the size of the family was an individual dual decision; it came about as an automatic consequence of an increasing standard of living of the masses. People began thinking that with a large number of children, they would not be in a position to attain—and maintain—a high standard of living. In countries like the United Kingdom, social legislation such as prohibition of child labour, compulsory primary education, etc., produced a situation under which children came to be considered a liability instead of an economic asset. Furthermore, as the status of women in society increased as a result of reforms such as the right to vote and increased job opportunities, small families became the accepted pattern.

India cannot wait either for a substantial increase in per capita income, or for a near universal education of the masses to bring about a decline in fertility. Both are likely to take a number of years, which India can ill-afford because during that period the population might more than double itself. As it is, the population is likely to reach a billion by the end of the present century. The Government, therefore, is depending upon a large-scale "motivational" campaign to bring about the necessary change in the attitude and values of the people. This is a new experiment in family planning strategy.

Family planning is a gigantic operation which requires the cooperation of medical doctors, social workers, demographers, economists, social scientists and administrators. While doctors are required for providing service and social workers for motivational work, demographers are required for evaluation and target-setting. The task of demographers, however, is difficult because of the nonavailability of reliable data for the general population, much less for population by subgroups. The need, therefore, for establishing "Population Laboratories" for obtaining more reliable and recent data, for perfecting techniques of accurate data collection, and for conducting methodological studies for estimating nonsampling errors, recall lapse, etc. is urgent.

There is also need to carry out research studies in depth; specially those dealing with attitudinal and motivational problems. Sociologists can help the family planning programme by indicating the factors which bring about social change and by ranking them, in order of importance so that they can be given priority. Medical doctors can give support to the family planning programme by inventing more acceptable methods of contraceptives. Economists could lend their techniques for undertaking meaningful cost- benefit studies so that the administrators can run the family planning programme at minimum cost and determine suitable priorities for expenditure. Family planning is multidisciplinary in nature and there is urgent need for different disciplines coming together on a common platform and giving of their best to solve this important problem.

Family Planning Progress During 1971-72

During the year 1971-72, about 2.19 million sterilisations were performed, nearly 0.49 million IUCD insertions were done and roughly 2.35 million couples were using conventional contraceptives, mainly Nirodh. In terms of achievement, the target for sterilisation during 1971-72 was more than fully achieved whereas for IUCD and conventional contraceptives about 60 per cent of the target was achieved. If the aggregate performance figures are studied, it can be said that since the beginning of the programme till March 31, 1972, a total of 10.86 million sterilisations were performed and 4.29 million IUCDs inserted. In addition, there were 2.35 million couples using conventional contraceptives during the financial year. Thus, up to March 1972 a total of 17.5 million couples were converted in favour of family planning. But all of them were not eligible couples as the number would deplete because of death or crossing of the reproductive period. Taking into account the above depletion factors, the number of couples currently protected during 1971-72 works out to 13.06 million or roughly 13.7 per cent of the eligible couples.

The use effectiveness of all the methods is 100 per cent. The use effectiveness of sterilisation and oral pills is 100 per cent, 95 per cent for IUCD and 50 per cent for conventional contraceptives. Thus, during 1971-72 a total of 11.82 million couples, or 12.4 per cent of the eligible couples, were effectively protected against conception. According to the estimates given by the Department of Family Welfare, 1.6 million births were averted during 1969-70; 1.9 million during 1970-71; and 2.1 million during 1971-72. These figures include the births which were averted due to the family planning work done in the previous years. During the decade 1961-71, around a total of 6.7 million births were averted due to family planning efforts, about 6.2 million since 1966-67. The birth rate which was around 41 in 1966 consequently came down to the level of 39 in 1971.

Because of the good performance in the previous year, the targets of family planning methods for 1972-73 were increased considerably, specially for sterilisation. Though the number of sterilisations during the year broke all previous records, in terms of achievement as a percentage of the target it dropped considerably. However, there was an increase in the percentage of eligible couples effectively protected and also in the number of births averted. During 1973-74, a steep decline in sterilisation was observed from 3.12 million in 1972-73 to 0.94 million in 1973-74. The fall was mainly due to the uncertainties arising out of the financial cuts in the budget. As regards IUCD and CC users, there was an increase compared to the previous year both in terms of number as well as achievement.

The percentage of eligible couples effectively protected increased slightly and a total of 3.0 million births were averted during the period due to various methods. During the year 1974-75, the first year of the Fifth Five Year Plan, the performance of sterilisation and IUCD was better than in the preceding year. The achievement as a percentage of the target also increased. However, in respect of the CC users there was a decline in the number of users which could be due to the steep fall in the sale of commercial Nirodh, the sale price of which was raised from 15 paise to 25 paise for a packet of three. However, there was a slight increase in the percentage of eligible couples effectively protected and during the year about 3.0 million births were averted.

During the year 1975-76, the targets of the family planning methods were increased. In terms of achievement, the target of sterilisation was more than fully achieved. Compared to the performance of the previous year, there was an increase in the ' achievement of CC users. However, there was a decline in the achievement of IUCD insertions. As a consequence of the good performance, specially for sterilisations, there was an increase in the percentage of couples effectively-protected. In the year 1974-75 the percentage of couples effectively protected was 15.0 which increased to 17.1 in

1975-76. Again, during the year a total of 3.13 million births were averted, thus, bringing the cumulative total of births averted to the level of 20.6 million. The year opened with the announcement of a National Population Policy which was the first of its kind in the history of Family Planning Programme in India. The policy statement outlined a series of measures to achieve a reduction in the birth rate to the level of 25 by the end of the Sixth Plan, The measures included certain incentives and disincentives aimed at promoting family planning as a mass movement.

The text of the policy statement is given elsewhere in this book. On the performance side the year 1976M recorded an extraordinary increase in the number of sterilisations. They increased to 8.26 million against 2.67 million in 1975-76. The achievement of sterilisation registered 192 per cent of the target. However, the performance of IUCD insertions and CC users was not so good; instead, compared to the previous year the achievement declined. Again, in terms of the percentage of couples effectively protected, the level increased to 23.7 per cent which was the maximum. The number of births averted during the year also increased to 3.7 million.

The President in his address to Parliament on March 28, 1977 stated that family planning will be pursued vigorously as a wholly voluntary programme and as an integral part of a comprehensive policy covering education, health, maternity and child care, family welfare, women's rights and nutrition. The full text of the address is given elsewhere. Again, during the same year the name of the programme was changed from family planning to family welfare which showed the Government's anxiety to promote the total welfare of the family and the community. Another important event which took place during the year was the passing of the

Child Marriage Restraint Act, 1978 which raised the minimum age of marriage from 15 to 18 years for girls and 18 to 21 years for boys. It may be mentioned the same was outlined in the National Population Policy of 1976. On the performance side, during the year 1977-78 there was a big slump in achievement. Though no target was fixed for sterilisation, the number of sterilisations performed during the year declined to less than one million.

Only one-third of the target of IUCD insertions was achieved which was the lowest. The achievement of the target of CC users declined compared to the previous year but was still more than 65 per cent. Because of the decline in the performance, the percentage of eligible couples effectively protected also registered a decline. However, because of the good performance in the previous year, the number of births averted during 1977-78 increased to more than 5.0 million which was highest in all the years till 1980-81. In the year 1978-79, there was some recovery in the performance of IUCD insertions and CC users.

The achievement in the targets of the two methods was about 92 per cent for IUCD and nearly 87 per cent for CC users. The number of sterilisations done during the year though they increased to around 1.5 million, but only 37 per cent of the target was achieved. As a result the number of births averted during the year and the percentage of couples effectively protected declined only slightly.

During the year 1979-80, there was further recovery in the number of sterilisations performed and IUCD insertions. The number of CC users registered a decline. However, the achievement of the targets of all the three methods was between 55 and 59 per cent only. In comparison to the previous year the levels of percentage couples effectively protected and the number of births averted during the year remained more or less the same. In the year 1980-81, there was significant improvement in the performance of the family planning programme. The-number of sterilisations performed during the year reached the 2 million mark and the achievement was 70 per cent of the target. In the case of IUCD and CC users the achievement was good.

As a consequence of good performance, there was an increase in the percentage of couples effectively protected. About 4.9 million births were averted during the year. It may be mentioned that the year 1980-81 was the first year of the Sixth Five Year Plan. During the period the Working Group on Population Policy submitted its recommendations to the Planning Commission and after the approval of the National Development Council the recommen-dations were incorporated in the Sixth Plan. The objectives of the Population Policy, which is sometimes known as the new population policy, as laid down in the Sixth Five Year Plan document is stated elsewhere.

A new Act called the Medical Termination of Pregnancy (MTP) Act came into force in India with effect from 1 April 1972. Under this Act, a pregnant woman above the age of 18 can get her pregnancy terminated within the first 12 weeks by a qualified doctor; the only condition is that in the opinion of the doctor the continuation of the pregnancy either involves a risk to the life of the mother or is likely to cause great injury to her mental and physical health, or if the child is likely to suffer from physical or mental abnormalities.

In case the duration of pregnancy exceeds 12 weeks but is less than 20 weeks, an abortion can be performed only when the opinion of two doctors is taken and both of them agree that the continuation of pregnancy will involve substantial risk to the health of the mother or is likely to result in the birth of a defective child. Pregnancy of a minor woman or of a lunatic woman can be terminated only after written consent from her guardian has been obtained. In all cases, however, the consent of the pregnant woman herself is necessary.

The trend of MTP cases shows that it is increasing except for the year 1977-78 when it slightly dropped compared to the previous year. During 1980-81, a total of 385,749 MTPs were performed as against 360,838 in 1979-80. During the year 1979-80, the largest number of abortions were carried out in Uttar Pradesh (24.6 per cent), followed by Maharashtra (10.9 per cent), Tamil Nadu and Kerala. An analysis based on the MTPs performed during 1977-78, 1978-79 and 1979-80 reveals that in 82-85 per cent of the cases, the abortions were performed within 12 weeks of pregnancy, in 77-79 per cent of the cases the mother's age was between 20 and 35 years. Again, in 46-48 per cent of the cases the abortions were performed due to failure of contraceptives, 14-16 per cent of the cases were due to grave injury to mental health and another 15-18 per cent of the cases due to grave injury to physical health.

The request for abortion had come from women of all religious groups. Of the cases reported about 85 per cent were Hindu, around 7.5 per cent Muslim, nearly 3.5 per cent Christian and about 2.5 per cent Sikhs. Abortions under the MTP programme are conducted by trained doctors in Government hospitals and approved private institutions. To get more trained personnel, doctors from various institutions, for example, medical colleges, district hospitals, PHCs, voluntary organisations and private practitioners are being trained. By the end of March 1981, a total of 7,037 doctors were trained out of which 1,847 were from PHCs under the programme. As on March 31 1981, 3,294 institutions were approved to perform the abortions under the programme. Again by the end of March 1981, slightly more than 1.97 million abortions were performed since the inception of the MTP programme in 1972. It has been mentioned earlier that the minimum age of female marriage has been raised from 15 years to 18. It is hoped that with this measure the length of ferthe union will be shortened by at least two years; and this will tend to reduce the birth rate. This measure is also expected to influence the fertility performance as girls marrying later will be more mature, and, hopefully, more amenable to the idea of family planning.

In the exercise of NRR = 1 by 1996, the marriage pattern has been considered as one of the inputs which represents the social change. As such a machinery should be created to educate the people and explain to them the benefits of raising the age at marriage/ marriages at the right time. The Department of Family Planning is also seriously thinking of raising the minimum age of female marriage from the present fifteen years to eighteen. It is hoped that by this measure females will shorten their effective fertility union by three years; this will tend to reduce the birth rate. The measure is also expected to influence the fertility performance of women as girls marrying later will be more mature, and, hopefully, more amenable to the idea of family planning.

The Department has prepared a draft bill and has sent it to various State Governments to elicit their opinion. Under the present Child Marriage Act an offence committed under this Act has to be brought to the notice of the Government, but people are reluctant to do so. The Government, therefore, is keen to create, a machinery to report the offence to overcome this defect. The machinery proposed to be created under this legislation will also be of an educative character and the officers appointed under this scheme will be expected to associate with the local people and explain to them the benefits of raising the age at marriage.

The status of women is known to have important implications for fertility performance. It is maintained that if the status of women rises, they tend have fewer children and they can thus make a meaningful contribution society. Also, if women are tied down to children bearing for the most fruitful ye avenues for self-expression is well protected today. Earlier, the population policy of India was equated with the family planning policy. A major departure from this approach was made when, on 26 April 1976, the Minister of Health and Family Planning, Dr. Karan Singh, declared the new National Population Policy of India. This policy statement takes into account the complex relationships between the social, economic, and political aspects of the population problem, and defines the way in which the family planning programme could be pushed ahead with due attention given to the various facets of the problem.

The text of the policy statement is reproduced here in full as it denotes an important development in the thinking of the Government of the country on this vital problem. "With 2.4 per cent of the world's land area, India has about 15 per cent of the world's people. It is estimated that our population as on I January 1976 has crossed the 600 million mark, and is now rising at the rate of well over one million per month. Since Independence 250 millions have been added, equivalent to the entire population of the Soviet Union with six times the land area of India.

The increase every year is now equal to the entire population of Australia which is two and a half times the size of our country. If the present rate of increase continues unchecked our population at the turn of the century may well reach the staggering figure of one billion. Indisputably, we are facing a population explosion of crisis dimensions which has largely diluted the fruits of the remarkable economic progress that we have made over the last two decades. If the future of the nation is to be secured, and the goal of removing poverty to be attained, the population problem will have to be treated as a top national priority and commitment.

Our real enemy is poverty, and it is as a frontal assault on the citadels of poverty that the Fifth Five-Year Plan has included the Minimum Needs Programme. One of its five items is an integrated package of health, family planning and nutrition. Far-reaching steps have been initiated to reorient the thrust of medical education so as to strengthen community medicine and rural health aspects, and to restructure the healthcare delivery system on a three-tier basis going down to the most far-flung rural areas where the majority of our people reside and where child mortality and morbidity are the highest. Similarly, ignorance, illiteracy and superstition have got to be fought and eliminated. In the ultimate analysis it is only when the underlying causes of poverty and disease are eliminated that the nation will be able to move forward to its desired ideals.

Nonetheless, it is clear that simply to wait for education and economic development to bring about a drop in fertility is not a practical solution. The very increase in population makes economic development slow and more difficult of achievement. The time factor is so pressing, and the population growth so formidable, that we have to get out of the vicious circle through a direct assault upon this problem as a national commitment. The President, in his address to Joint Session of Parliament this year, reiterated the importance of stepping up family planning efforts, and the Prime Minister has on several occasions laid stress upon the crucial role that population control has to play in the movement towards economic independence and social transformation, specially in the light of the 20-Point Economic Programme. Considerable work has been done in our country in the field of family planning, but clearly only the fringe of the problem has so far been touched. In this context, after a thorough and careful consideration of all the factors involved as well as the expression of a wide spectrum of public opinion, Government have decided on a series of fundamental measures detailed below which, it is hoped, will enable us to achieve the planned target of reducing the birth rate from an estimated 35 per thousand in the beginning of the Fifth Plan to 25 per thousand at the end of the Sixth.

Allowing for the steady decline in the death-rate that will continue due to the improvement in our medical and public health services and the living standards of our people, this is expected to bring down the growth rate of population in our country to 1.4 per cent by 1984. Raising the age of marriage will not only have a demonstrable demographic impact, but will also lead to more responsible parenthood and help to safeguard the health of the mother and the child. It is well known that very early pregnancy leads to higher maternal and infant mortality. Also, if the women of our country are to play their rightful role in its economic, social and intellectual life, the practice of early marriage will have to be severely discouraged.

The present law has not been effectively or uniformly enforced. It has, therefore, been decided that the minimum age of marriage should be raised to 18 for girls and 21 for boys, and suitable legislation to this effect will be passed. Offences under this law will be cognisable by an officer not below the rank of a Sub-Divisional Magistrate. The question of making registration of marriages compulsory is under active consideration. It has been represented by some States that while on the one hand we are urging them to limit their population, those States which do well in this field face reduction of representation in Parliament while those with weak performance in family planning tend to get increasing representation. It is obviously necessary to remedy this situation. It has, therefore, been decided that the representation in the Lok Sabha and the State Legislatures will be frozen on the basis of the 1971 Census until the year 2001. This means in effect that the census counts of 1981 and 1991 will not be considered for purposes of adjustment of Lok Sabha Legislature seats. Necessary constitutional amendment will be brought forward during the current year. Appropriate legislation for other elective bodies will also be undertaken.

In a federal system, the sharing of Central resources with the States is a matter of considerable importance. In all cases where population is a factor, as in the allocation of Central assistance to State Plans, devolution of taxes and duties and grants-in-aid, the population figures of 1971 will continue to be followed till the year 2001. In the matter of Central assistance to State Plans, eight per cent will be specifically earmarked against performance in family planning. The detailed procedures in this regard will be worked out by the Planning Commission.

While there is a direct correlation between illiteracy and fertility, this is particularly marked in the case of girls' education. Wherever female literacy improves, it has been seen that fertility drops almost automatically. It is, therefore, necessary that special measures be taken to raise the levels of female education, particularly above the middle level for girls as well as nonformal education plans for young women especially in certain backward States where the family planning performance so far has been unimpressive. The same is true with regard to child nutrition programmes, as high infant mortality and morbidity have a direct impact on fertility.

The Ministry of Education is urging upon the State Governments the necessity to give these matters higher priority than has been accorded so far and fully earmarking adequate outlays both for girls' education up to the middle level and child nutrition. It is essential that the younger generation should grow up with an adequate awareness of the population problem and a realisation of their national responsibility in this regard.

The adoption of a small family norm is too important a matter to be considered the responsibility of only one Ministry. It is essential that all Ministries and Departments of the Government of India as well as the States should take up, as an integral part of their normal programme and budgets, the motivation of citizens to adopt responsible reproductive behaviour both in their own as well as the national interest. A directive to this effect is being issued by the Prime Minister to all Ministries of the Government of India, and a letter will also be addressed by her to all Chief Ministers.

The performance of family planning in the States will be more carefully and intensively monitored than in the past, and the Union Cabinet will review the situation in depth at least once a year. Experience over the last twenty years has shown that monetary compensation does have a significant impact upon the acceptance of family planning, particularly among the poorer sections of society. In view of the desirability of limiting the family size to two or three it has been decided that monetary compensation for sterilisation will be raised to Rs. 1501 if performed with two living children or less, Rs. 100 if performed with three living children and Rs. 70 if performed with four or more children. These amounts will include the money payable to individual acceptors as well as other charges such as drugs and dressing, etc. and will take effect from 1 May 1976. Facilities for sterilisation and MTP are being increasingly extended to cover rural areas. In addition to individual compensation, Government is of the view that group incentives should now be introduced in a bold and imaginative manner so as to make family planning a mass movement with greater community involvement. It has, therefore, been decided that suitable group incentives will be introduced for the medical profession, for Zila and Panchayat Samities, for teachers at various levels, 'for cooperative societies and for labour in the organised sector through their respective representative national organisations. Details of these group incentives are being worked out in consultation with the concerned organisations.

Despite Governmental efforts at Union, State and Municipal levels, family planning cannot' succeed unless voluntary organisations are drawn into its promotion in an increasing measure, particularly youth and women's organisations. There is already a scheme for aiding voluntary organisations, and it has been decided that this will be expanded. Also, full rebate will be allowed in the income tax assessment for amounts given as donations for family planning purposes to Government, local bodies or any registered voluntary organisation approved for this purpose by the Union Ministry of Health.

Research in reproductive biology and contraception is under way in several of our scientific institutions, and there are some very promising developments which, we hope, will lead to a major breakthrough before too long. This is a greet challenge to our scientists, and efforts in this direction will receive special attention so that necessary research inputs are ensured on a long-range and continuing basis.

The question of compulsory sterilisation has been the subject of lively public debate over the last few months. It is clear that public opinion is now ready to accept much more stringent measures for family planning than before. However, the administrative and medical infrastructure in many parts of the country is still not adequate to cope with the vast implications of nationwide compulsory sterilisation. We do not, therefore, intend to bring in central legislation for this purpose, at least for the time being. Some States feel that the facilities available with them are adequate to meet the requirements of compulsory sterilisation. We are of the view that where a state leislature, in the exercise of its own powers, decides that the time is ripe and it is necessary to pass legislation for compulsory sterilisation, it may do so.

Our advice to the States in such cases will be to bring in the limitation after three children, and to make it uniformly applicable to all Indian citizens resident in that State without distinction of caste, creed or community. Some States have also introduced a series of measures directed towards their employees and other citizens in the matter of preferential allotment of houses, loans, etc., for those who have accepted family planning. In this sphere also we have decided to leave it to each individual State to introduce such measures as they consider necessary and desirable. Employees of the Union Government will be expected to adopt the small family norm and necessary changes will be made in their, service/conduct rules to ensure this.

In order to spread the message of family planning throughout the nation, a new multimedia motivational strategy is being evolved which will utilise all the available media channels including the radio, television, the press, films, visual displays and also include traditional folk media such as the jatra, puppet shows, folk songs and folk dances. The attempt is to move from the some what urban elitist approaches of the past into a much more imaginative and vigorous rural-oriented approach. In this context my Ministry is working in close coordination with the Ministry of Information and Broadcasting, and is also trying to draw the best media talent available in the country into the structuring of the new programme. This package of measures will succeed in its objective only if it receives the full and active cooperation of the people at large. It is my sincere hope that the entire nation will strongly endorse the new population policy which, as part of a multifaceted strategy for economic development and social emancipation, is directed towards building a strong and prosperous India in the years and decades to come.

On the recommendations of the Working Group on Population Policy, the Government declared the objectives of the Family Welfare Programme. The objectives of the Family Welfare Programme as given in the Sixth Five Year Plan document are given below. The Working Group on Population Policy set up by the Planning Commission has recommended the adoption of the long-term demographic goal of reducing the net reproduction rate (NRR) to one by 1966 for the country as a whole and by 2001 in all the States from the present level of 1.67.

The implications of this are as follows: (i) The average size of the family would be reduced from 4.2 children to 2.3 children. (ii) The birth rate per thousand population would be reduced from the level of 33 in 1978 to 21. (iii) The death rate per thousand population would be reduced from about 14 in 1978 to 9 and the infant mortality rate would be reduced from 129 to 60 or less. (iv) As against 22 per cent of the eligible couples protected with family planning at present 60 per cent would be protected. (v) The population of India will be around 900 by the turn of the century and will stabilise at 1200 million by the year A.D. 2050. Keeping in view the long-term demographic goal of reducing the NRR to I by 1995 (sic) as approved by the National Development Council, the following targets have been envisaged for the Sixth Plan keeping in view the past performance, present capacity and future potential: The number of sterilisations which were around 1.74 million in the base year (1970-80) will go up to 5.50 million in the terminal year of the Plan.

The number of IUD insertions will go up from 0.62 million in 1979-80 to 2.50 million in 1985. The percentage of effective couple protection envisaged by 1985 is 36.56 against the present 22.5. The calls for a tremendous motivational effort matched by adequate infrastructure. In this context mention must be made of a total of 15.5 million sterilisations that will be required during the plan period 1980-85 for maintaining the birth rate at the existing level, assuming the present levels of IUD and CC users.

7 Future of Man

The most difficult and essential problems for man's future lie in broad ecologic and social issues. No two questions are more fundamental or more troublesome to the future of man than the ecologists' inquiry. "What is the carrying capacity of the earth for man? In other words, what are the limits of the life support capabilities of the earth in terms of human population?" are the behaviorists' concern. "What are the tolerance levels of man to the pressures generated by his own societies? In other words, if man is his own worst enemy, does this lead to his self destruction?"

The carrying capacity, or life support capability, of the earth for man may be dependent upon a great variety of factors. We can readily identify oxygen water, food space and shelter as vital requirements for human life. There might be many others as well such as mineral resources, energy availability, or the concentration of toxic metabolites, which could reach states above human tolerance levels. Most of these however, would operate through the media of air, water or food. Hence, it is worth while to examine each of these briefly.

The atmosphere of the earth consists of about 78 percent nitrogen, 21 percent oxygen, 0.03 percent carbon dioxide, and the remainder, other gases. This composition is remarkably stable despite the fact that oxygen is a highly reactive element which readily combines with a great many other elements in chemical processes of oxidation. The ecologist Lamont C Cole (1970) has pointed out that oxygen has remained stable in the earth's atmosphere because green plants recycle it in molecular form through photosynthetic process. Approximately 70 percent of the free oxygen produced each year is through to come from planktonic plants in the ocean, and the remaining 30 percent from terrestrial vegetation of forests and grasslands.

Cole emphasized that if we should seriously pollute the sea, as is already happening in coastal areas, and if we should excessively denude the forests and grasslands of the earth, we shall be in danger of upsetting the oxygen balance in the atmosphere. In other words, we must look upon the oceans, forests and grasslands, as the world's oxygen tank, and we must insure their continued healthy ecologic function. This point of view is contested by W.S.Broecker (1970), a geochemist at Columbia University Dr. Broecker points out that each square meter of the earth's surface is convered by 60,000 moles of oxygen gas. Plants in the ocean and land produce only about 8 moles per square meter per year, an this is almost entirely consumed by their own respiration and that of animals and bacteria.

Thus other organisms use all but 0.0004 percent of the oxygen produced by plants. Broecker maintains, therefore, that man does not depend upon the oxygen generated by plants, and that there is no current danger at all to his oxygen supply. "In conclusion it can be stated with some confidence that the molecular oxygen supply in the atmosphere and the broad expanse of open ocean are not threatened by man's activities in the foreseeable future. Molecular oxygen is one resource that is virtually unlimited". Broecker feels that man's existence will be threatened by other types of environmental pollution long before his oxygen supply becomes endangered, and that we need not worry about total oxygen supply. This is typical of some of the different opinions which are unfolding between physical and biological scientists. Such differences often indicate the need for more research and better knowledge before we can fully accept one position on another.

Another atmospheric problem of long-term concern is the increase in carbon dioxide in the atmosphere. Between 1860 and 1960 the combustion of fuel increased the CO_2 content of the air by 14 percent of its formerly stable level (Commoner, 1966), Carbon dioxide tends to reflect infrared rays from the earth, and this provides a "greenhouse effect" for the whole earth; that is, it accumulates heat. This will gradually increase the temperature of the earth—an effect which may have long-term implications in climate modification.

A rise of just 2 or 3 degrees in annual mean temperature would have a major melting effect on glaciers and ice caps, and this could raise the level of the seas several feet. Such a rise in the seal level would inundate many of the world's major cities. Some experts predict it may require 4,000 years, and others feel it will not happen at all. The latter believe that any occurrence of the greenhouse effect due to increasing CO_2 in the atmosphere will be offset by increasing particulate pollution in the air and a cooling of the earth's surface. Once gain, we see fundamental controversies on major issues-largely because we do not really have the complete answer.

The atmospheric problem may also become acute in terms of various toxic elements released into surface, air and water: carbon monoxide, sulfur dioxide, lead from automobile exhaust, hydrocarbons from industrial operations and numerous other products of modern technological societies. For example, 500 million pounds of lead are discharged into the atmosphere of the United States each year from automobile exhausts alone, and significant amounts of this become incorporated into living organisms.

A recent study in Sweden showed concentrations of lead in mosses to be as high as 300 to 500 ppm along roadsides. Some roadside plants in the United States may have lead levels up to 3,000 ppm,. This lead enters the plants via both air and water and can then enter man and animals. A hundred years ago, before the automobile, concentrations in roadside plants were less than 20 ppm. Lead levels over 0.5 ppm have detectable physiological effects in man and animals, and 1.0 ppm levels are often toxic.

There seems to be little or no immediate threat that we will run out of air for breathing, but several warnings have appeared regarding the quality of our air. We apparently have an adequate supply of total oxygen for many centuries to come though even this is a debatable point. In some localities the situation may be critical at certain times. If stationary inversions of air trap the polluted atmosphere over some of our largest cities such as Tokyo, London, New York or Los Angels, toxic air pollutants increase to dangerous levels. In August of 1969, a massive still-air inversion occurred in eastern United States, causing serious air pollution in urban areas from New Orleans and St. Louis, north to Chicago and Milwaukee, and east to Washington and Baltimore.

At no time was there an absolute shortage of oxygen, but dangerous level of carbon monoxide, sulfur dioxide, hydrocarbons and other air pollutants accumulated in several cities. Twenty million people were affected by varying degrees of eye irritation and respiratory diseases. Fortunately, this great blanket of smog was dispersed by upper air currents in the first week of September, and no large-scale health tragedies occurred. More serious event of this type are likely to occur in urban areas unless strong measures against air pollution are undertaken.

Worldwide water supplies represent a concern for the future of man in terms of both the quality and absolute quantity of fresh water. Many cities have reached or are coming close to their total available fresh water supply and are reaching further and further to obtain new supplies. Los Angeles, for example, is planning tremendous canals and canduits which will bring water from the pacific northwest over 1,000 miles to the growing megalopolis of southern California. Two Russian geographers and hydrologists, Kalinin and Bykov, have pointed out that "The world's water resources from a single entity.

The basic hydrologic cycle of the world depends upon solar energy, atmospheric circulation, heat reflectivity and surface retention. The latter is strongly influenced by the biotic community, kalinin and Bykov point out, on the assumption that the moisture content of the earth remains constant that the total gains and losses from the land and the sea are in balance. Thus, the total precipitation equals the total evaporative loss from land and sea. However, the relative retention time and thus the total amount of water in terrestrial and fresh-water environments can change substantially in relation to the biota and terrestrial storage capacity.

A large proportion of the world's fresh water, falls and resides in relatively unpopulated portions of the globe, particularly Siberia and northern Canada. Here, man's activities are limited by other factors and he cannot readily obtain these vast stores of water. Secondly, the matter of water quality is of immediate concern. The principle of water reuse is now becoming accepted, so that people recognize much of their water has been used previously by other municipal systems. Each reuse cycle however, is increasingly costly in terms of treatment, and each reuse cycle requires disinfectants and other industrial chemicals, combining these problems of distribution and quality, it seems that the water problem of the future is not going to be one of a global water shortage on a planetary basis but one of local crises due to the concentrations of people, industry and water use.

The immediate concern is finding local water resources of adequate quality to meet the burgeoning needs of urban populations. To the city dweller in New York or Hong Kong whose water tap runs dry or to the refugee in the Middle East who is dying of thirst, it well be little comfort to realise that 29,000,000 cubic kilometers of fresh water are locked in glaciers and polar ice caps.

Nowhere is scientific controversy more evident than on the issue of world food supplies. Some scientists predict famines and mass starvation within the next 10 years. They feel the stage is irrevocably set for millions of people to die of starvation in the 1980's. In fact, we already have famine in many parts of the world, if one defines famine as people dying of lack of food. Even at the height of the Green Revolution in Asia in which new strains of rice and wheat were producing phenomenally large yields, famine conditions existed in parts of Rajasthan in India: in sections of Burma and in southern Thailand. Starvation is now a grim reality in much of the world and it may well get worse despite the Green revolution.

Other scientists refute the pessimism and find evidence for a hopeful outlook in the world food situation. They feel that the Green Revolution which was based on new strains of rice and wheat in Asia and Africa, and which produced a doubling and tripling of yields in one or two years, shows what can be done with modern agriculture. For the first time in many decades, agricultural output increased more rapidly than population growth. India began to approach self-sufficiency in food production, and the Philippines began to export rice again for the first time since 1900.

A word of caution about the Green revolution is also necessary. It is quite true that new strains of agricultural crops can sometimes dramatically increase yields, but often these new strains require more careful management, more intensive care, more water, more fertilizer and more insecticides. They may be more vulnerable to plant viruses or fungi. Although the plant geneticists try to breed in every possible disease resistance, these new strains do not have the centuries of natural selection and regional adaptation behind them as do the local strains which the villages have always used. In West Bengal and Orissa, there were reports that some of the newly introduced high-yield strains of rice were succumbing to a disease that was not affecting that old traditional strains. If this happens to a major extent, the Green Revolution could collapse and a severe blow would be dealt to the villager's faith in the agricultural scientist.

Another factor in the world food picture is one of pure climatic good fortune. Green Revolution in India was accompanied by the best monsoon seasons in over 10 years. A long series of droughts had broken is southern Asia and adequate rains fell in the right pattern. The possibility exists, therefore, that a good part of the success of the Green Revolution was due to good weather, combined with new strains and more fertilizer. This is a question that will be more adequately answered in future year to see if increasing agricultural production can be maintained in years of poor rainfall. In India too much rain came and it fell to early. The wheat crops in northern India were magnificent in April and May just before harvesting and threshing, but heavy premature rains in many areas caused the wheat to rot before it could be threshed. Bihar had serious crop losses from flooding. Once again, this was a tragic and dramatic demonstration that even the Green Revolution remains dependent on climatic factors outside the control of man.

Through all of these arguments on how many people the world can feed, one must ask the realistic question: "Is it necessary that there always be a race between population and food production? Is it desirable to be teetering on the brink of starvation at all times? Is it the best representation of the human condition to be struggling in this manner to feed more and more people at a subsistence level? One also has the impression in reviewing the literature on world food supplies that too much emphasis is placed on future predictions, and not enough on current conditions. We furthermore know that malnutrition exist throughout the world, even in the most advanced countries.

We furthermore know that malnutrition of many kind has profound effects upon growth, development and behavior. Children growing up with malnutrition from early infancy definitely show reduced growth rates and possible retardation of physical and mental development. They furthermore show increased susceptibility to infectious disease. Thus, we return to the issue of the quality of life, and the realization that merely preventing death by starvation does not by any means insure a quality nutritional base for adequate growth and human development. It is apparent that the problem of world food supplies will continue for some time as a problem of both quantity, and quality and especially one of distribution.

In 1958 Paul Sears wrote an important essay entitled, "The Inexorable Prooblem of Space," in which he concluded that "Our future security may depend less upon priority in exploring outer space that upon our wisdom in managing the space in which we live". The earth's total surface is 196,940,400 square miles of which about 30 per cent or 57,230,000 square miles is land surface. Of this total land surface, approximately 5,000,000 square miles are in Antarctica deserts, and 11,000,000 square miles are in mountainous regions. Of deserts and moutains, at least one-fourth of their jano area cannot be inhabited, so this eliminates another 5,000,000 square miles of land surface for human habitation. We are left with approximately 47,000,000 square miles of land surface, at the most, for the world's population.

The world now has nearly about 6 billion people. The world population has doubled in the past 40 years and may double again in the next century, perhaps approaching stability at about 11 billion by the year 2100. By 1990, of the world's 5.3 billion people, 4.1 billion—77 percent lived in the developing world; 1.2 billion inhabited the industrialized countries. India has nearly 1/6th of the world's population. There are 257 person per sq. km. in India. In the Capital City Delhi there are 6318 person per sq. km. Thus it is to say human population is not evenly distributed. It leads to the problem of space.

As with water, air and food, it is important to distinguish between total or potential amount of any given resource on earth and that amount which is actually utilized or practically available to man, it was pointed out in the sections on water and air that the earth's total amount is apparently adequate for many years to come, but the supply in any given city may not be of sufficient quantity or quality to support the standard of living we would like. The amount of living space per person in some cities is less than 20 square feet per person. Theoretically, we could redistribute these crowded people to provide more even spacing around the world. Obviously, this cannot be done, and if it were attempted it might wreak havoc with agricultural production. The economic and behavioral needs of man are such that he must cluster together, but in so doing, he creates superdensities such as Calcutta, London, New York, Paris, Tokyo, Shanghai and Buenos Aires.

The slums of which are definitely deleterious to human health and welfare. Why, we might ask, does man insist on this degree of crowding and then complain about it so bitterly? Why do people continue to pour into the cities of the world when they are already known as centers of population crime and illness? Perhaps we need to know more about the real spatial and social needs of man. What are his requirements for space and privacy, and how do these interact with his needs for social stimulation and group activity? Does man become conditioned to crowding in early childhood and then seek further crowding throughout his life? Does he require the social pressure of a crowd because he feels a compulsion to be in touch with the group and thus be able to predict what the group is going to do next? Have we produced, in our urban centers, self-destructive, systems which so modify the behavior of their inhabitants that these inhabitants exaggerate their own problems and can no longer achieve rational solutions?

These are all tough question that cannot be answered in simple or easy terms. The answers will require the best intelligence we can muster from many disciplines. Some sorts of answers will certainly be necessary before we can even begin to solve the ecologic and behavioral problems associated with our own use of terrestrial space.

With all of the last four life-support requirements of man-air, water, food and space—we have seen that the major problem of immediate importance lie in distribution more than in total amount. Technologist therefore look upon these problems as soluble by technological means. They feel, they are essentially matters of economics and transportation Behavioral and environmental scientists see them as more basic, however, relating to man's fundamental ecologic behavior. They feel, they are essentially matters of economics and transportation.

Behavioral and environmental scientists see them as more basic, however, relating to man's fundamental ecologic behavior. They feel that certain basic social and ecologic tendencies of man prevent equitable distributions of the earth's resources. Obviously this type of argument can go on forever, and it is perhaps necessary to take a pragmatic approach and look at the current picture for the best glimpse of the future. Many of the industrial nations of the world, especially the United States, the national northern Europe and Japan, have enjoyed rising prosperity for the last 25 years. In ecologic time, of course, this is but a flash in the pan, but let us assume that it can be sustained for another 25 or 50 years. It is worth examining where the material and energetic support for this prosperity arises.

This distributional inequality is another major symptom of ecologic instability in the world. One would hope for some narrowing in the gap between the haves and the have-nots, but unfortunately the opposite trend is still occurring. The wealthy nations of the world are far outstripping the poor nations in their rates of economic gain and human betterment. There are, however, self-corrective mechanisms within this system as, for example, when some segment of the labour it can find which will do a satisfactory job. Thus, there has been a great expansion of manufacturing in Japan, Hong Kong, Taiwan and Singapore in industries as diverse as steel, electronics, textiles, toys and optics. The Western factory worker experiences for the first time an improving situation.

The western industrial worker continues to refuse his own working prospects by striking for higher and higher wages. This will certainly be one mechanism by which wealth is more adequately balanced between nations. Still the mass of poverty throughout much of Asia, Africa and Latin America is so great that it is difficult to imagine a painless equilibrium. The differential distribution of wealth remains sharp and represents a type of ecologic situation which cannot last indefinitely without severe consequences and reverberations. To the ecologist it means either a successional stage which is not yet mature, or a system headed for a major breakdown.

Economic and ecologic maturity should ideally be in the direction of broader and more equitable distribution of the earth's resources—a narrowing of the gap between the haves and the have-nots, both within and between nations, provided the total system does not pass beyond the state of corrective change.

Any ecological system or biotic community can be overloaded to the point of collapse by a variety of factors. Certainly one of the factors for man of most immediate and realistic danger is that of social violence. Either by war or by civil riot, man threatens his own existence. To maintain a sense of physical security man has built an incredible stockpile of destructive power. The nations of the world spend more than 200 billion dollars per year on armaments—more than on health education, welfare, agriculture, child care and care for the aged combined. Twenty tons of explosives are stored in the world for every man, women and child. Russia and the United

States have stockpiled 6,000 nuclear war heads possessing thousands of megatons of explosive force. Each megaton has more destructive power than which existed in any of the previous wars of history. The world spends almost 8,000 dollars per year training each soldier and less than 100 dollars per year educating each child. The question is, of course, how long can man continue to preside over armament like this without using it? And the answer seems to be, not very long.

The fact is we do use part of the power in an attempt to achieve social goals. It is being used daily in various parts of the world to maintain territory and dominance ranks. It is justified as a spare and deliberate use in hopes of avoiding the all out effort. How long this can continue is anyone's guess, and the ecologist can claim no more omniscience than the political scientist. The ecologist points out, however, that the ecologic and social pressures which generate international military tension may rise with increased population pressure. It is not likely that man can solve the problems of war and violence until he approaches and understands their basic orgins in terms of the behavioral and ecologic relationship between social groups, and the environment.

Many scholars feel that one of the great problems of the present and the future is the rapid rate of change which is occurring in society and technology. There is not, however, agreement on the consequences of the rapid change, changing life patterns are too rapid for man to adapt successfully. Mankind is so overstimulated by the increase in new sight and sounds, so bombarded by world wide mass media, and so constantly beset with noval situations and new problems that he will break down psychologically and physically. His responses will be maladaptive and pathologic. He will thus collapse under sensory overload. It is quite possible that each of these points of view has large elements of truth in it. In the present world one can find many examples of both. Our mental institutions are already filled with those who could not adapt to sensory overload and who experience a personal breakdown of physical and mental health.

Yet one can find people are living a sub-human existence where the quality of life is so low that the amount of human pain and misery seem unbearable. Yet man has adapted successfully to these inhuman conditions. In many ways, the future is with us now. We can find throughout the world an abundance of high quality life in high quality environments, but we can also find an abundance of low quality life in impoverished environments. Many of the questions about the future revolve, therefore, around the question of relative balance and proportions. Which way are we heading and which way will the scales tip?

A major aspect of the future centers around our sense of responsibility to posterity. No representative of posterity is here to speak for the segment of society. How do the present populations of mankind regard their obligations to future generations? Are we willing to be more conservative in our life requirements, and are we, in fact, willing to reduce our rates of personal economic growth to insure a better environment and a greater abundance of resources for future generations? This may be one of the central issues of the next twenty years, and one on which the quality of present and future environments depends.

World-Wide Environmental Issues

Environmental problems have become a global issue: Some of the world wide issues are briefly described below:

(a) Climatic changes/Green house effect

Many studies show that the atmospheric pollution is on a global scale. Today, however, it is not only the air over our cities that is polluted, the entire atmosphere of our planet is now afflicted to some degree. Meteorologists talk about a nebulous veil of air pollution encircling the entire earth. Smong has been observed over the North Pole, and in their unlikely places.

There are range of environmental pollution issues originating largely in developed countries which are also of global significance. Carbon dioxide (CO_2) build-up leading to climate change is one of those key issues. Recent research indicates that increasing concentrations of CO_2 in the atmosphere, due mainly to the burning of fossil fuels could cause a warming trend leading to climate change in the next century. While there is still scientific uncertainty about the pace and impact of such a change it could be of sufficient magnitude to produce major physical, economic and social dislocations on a global scale.

A special challenge posed by the issue is that if preventive actions are postponed until some effects are perceived, it may be too late to avoid appreciable climate change. It is therefore in the interests of all nations to support international efforts to understand the phenomenon at early enough stage for action to be taken to reduce or avoid major economic, social and environmental disruptions around the world. However, as developed countries are currently responsible for over two thirds of total world emissions of CO_2 by fossil fuel combustion, they have a special responsibility for supporting and contributing to accelerated efforts to assess more precisely the nature of the problems and the alternatives for action to reduce or avoid any serious risks.

Another issue is the world-wide distribution and use of harmful or potentially harmful chemicals. There are now some 70,000 chemicals on the commercial market, and many are currently used and released into the environment with little or no knowledge of their potential long-term effects. It is estimated that another 1,000 new chemicals enter the market each year. Chemicals cross national boundaries in a variety of ways: they are transported by winds and water.

(b) Risks to Ozone Layer

Ozone is a different form of life sustaining oxygen, having three atoms instead of the usual two. But this third atom, scientists have found, is very unstable and can be eliminated by chlorine ions. The scientist have now found a close link between the family foam mattress, refrigerator and the air conditioning system of an average home and Ozone layer, 50 to 70 kilometers above the earth. They believe that free chlorine ions into the atmosphere, from plastics and other chemicals remove that extra atom of oxygen which makes Ozone. Only, in this case, the loss of the one atom changes the ability of the layer to act as a strainer to absorb the damaging ultraviolet B wavelengths from the sunlight.

The findings by the experts have revealed that the rate of Ozone Layer depletion is sufficiently large to present serious consequences for man, animals and plants. In addition, new threats to the ozone layer are being recognized. Take, for example, the refrigeration industry. Food preservation has become a necessity and refrigeration is here to stay. Only advanced technology can provide the solution to prevent the release of the damaging ions.

Another example is nitrogenous fertilizer which, it is suspected, release chemicals damaging to the Ozone layer. The demand for more food and the booming green revolution in many developing countries have pushed up the use of this fertilizer. Today the world uses about 50 megatons of it. In another 20 years the rate of uses will be 100-200 megatons.

Though the Ozone Layer accounts for less than one millionth part of the earth's atmosphere, it is of vital importance to the earth's life form. It has two main functions: first it absorbs the UV-B while allowing the heat generating infra-red rays to reach the earth freely. The heat from infrared rays sustains plants and human lives by keeping the temperature of the planet at a desired level. Secondly, it influences the whether by stabilizing the stratosphere which acts as a cap to the turbulent weather system in the troposphere. Scientists knowledge of the Ozone layer is still limited for want of advanced technology.

The presence of other substances in the stratosphere, which absorbs UV-B, makes it difficult to come to a numerical and logical conclusion. But the scientists do agree that an average 70 per cent of the total UV-B reaching this planet falls on the equatorial regions between the latitudes 30° S and 30° N. They also agree that for genetic reasons people with light eyes and light skin (including albinos) are most affected by UV-B.

Scientists today claim UV-B can destroy micro-organisms outright as well as cells in plants and animals. It also destroys the DN molecules and the genetic cell composition. Under natural conditions plants and animals have developed behavioural adaptation but gradual or sudden increase in UV radiation due to changes in the Ozone Layer is likely to disturb this balance. While the scientists agree that UV-B contributes to the induction of cancer they agree the dose response relationships are too complex and unknown. These make the quantitative analysis speculative, as regards the cause of melanoma cancer and resulting deaths.

The North European Conference of Scientists November, 1980, warned that non-melanoma skin cancer in light-skinned people would increase and many crops and many acquatic organisms, including shrimps ad fish eggs, could be damaged by increased UV-B infiltration. This could even affect the reproductive process of plants and animals. It recommended that existing satellite and ground based ozone observing system should be integrated through Global Ozone Observing System (GOOS). They also recommended updating of equipment and setting up of more stations for observing the ozone system. The scientists concerned with outer limits of the biosphere felt they are not yet technically equipped to measure accurately the rate of ozone depletion-already depleted by one percent. Though the danger is known, science has not yet provided a perfect model of measuring the depletion rate.

(c) Acid Rain

Every one realises how important the rain is. Without it the continents would be barren. Yet now the rain in parts of the earth has taken on a new and threatening complexity. It mixes in the air with pollution from burning fossil fuels particularly in power stations, factories and motor vehicles and brings down dilute sulphuric acid. This is killing fish and water life, and corroding buildings, including some of the world's most important ancient monuments. It may also damage forest and croplands, and possibly pose a substantial threat to health. Acid rain is now recognized as an international problem.

The air of industrial towns partly is cleared by chimneys at factories which push the pollution high into the air. These chimneys have made the things better locally by dispersing the pollutants, but aggravated the international difficulties. For the sulphur and nitrogen compounds emitted by burning fossil fuels can be blown thousands of kilometers by the winds, to cause acid rain in countries far from their points of origin. A special conference on acidification of the environment was held in Stockholm in 1982. It reviewed and assessed a large amount of scientific information not previously available. So far, acid rain has been seen as an essentially regional problem, confined to the industrial areas of the northern hemisphere. But, though the problem was first perceived there, it may become much more widespread for acid rain is likely to occur wherever fossil fuels are intensively used.

Acidification is an environmental problem, or becoming one, in parts of Europe and North America. Similarly polluted areas are likely to exist elsewhere in the world, especially around large urban and industrial coagulumerations. Soils are normally much better able to resist acidification than lakes, rivers and streams. Meanwhile, acid accelerates corrosion in most materials used in construction of historic buildings, and other important cultural objects.

Some of th worlds greatest cultural treasures including Parthenon in Athens and Trojans' column in Rome are being eaten away by acid fall out. There must be more effort to research and develop ways of improving energy conservation, environmentally appropriate technologies for producing heat and power, and techniques for removing sulphur from fossil fuels and gaseous emissions. The problem of acidification may well spread to new areas as a result of rapid industrialization and the growth of cities in other parts of the world, particularly developing countries. So it is important that areas affected by acid deposition and susceptible to damage from acidification are identified as soon as possible. It they are, the damage could be mitigated or even avoided of a minimum cost to society by initiating research and applying what we already know about pollution control.

(d) Pollution of World's Oceans

The data collected from the Baltic, the North Sea, and

some North American and Australian stuaries and coastal waters suggest that in most of these areas the pollution that caused most concern was due to sewage, agricultural chemicals, oil and metals. Metal concentrations were clearly elevated in coastal waters, and in fish and shelfish living there. In some areas, mercury levels in species such as tuna were high enough to make these fish unsuitable as human food, (Perhaps the most notorious incident of all is the outbreak of "Minamata Disease" in Japan. Methylmercury discharged from a chemical factory into the sea, or produced in the sea from inorgnic mercury discharges, contaminated the fish eaten by local people at the town of Minamata of Kyushu island, Japan. As a result of this and a similar incident at Nilgata on the east coast of Honshu, nearly two thousand people have so far suffered neurological disorders; about four hundred have died).

Some of the pollution came via river; the amount of iron, manganese, copper, zinc, lead, tin and antimony that reached the sea by this route was far greater than would be supplied by natural geological processes. Other contaminants came through atmospheric deposition: the importance of this pathway for metals and synthetic chemicals was increasingly recognized during the decade (1972-82). Offshore oil and gas exploration and dredging for sand and gravel in coastal areas increased during the period 1972-1982. Coastal zone development affected extensive estuarine areas, as well as mangrove swamps and coral reefs. Oil pollution killed sea-birds fouled beaches and affected tourism. Although tanker accidents were the source of less than 5 percent of all the oil entering the sea, accidents released large volumes in small areas, and were therefore especially damaging.

A recent report issued by a group of experts on marine pollution and published by the United Nations Environmental Programme, Warns that increasing contamination of the coastal seas, which has so far only caused localized disturbances to the marine environment may become much more serious. The combined effects of the dumping of sewage sludge industrial wastes, litter and oil spills into the Mediterranean, the North Sea, the Baltic and the Gulf of Mexico, may well cause ecological disasters which could gradually lead to a global crisis since the oceans are a unified life support system.

The report points out that the sea's capacity to absorb wastes has its limits, so that as the world economy continues to grow, even more sewage, industrial wastes and other effluents are likely to find their way into the sea. Man-made radioactive materials have been discovered in the open ocean and scientists have found traces of DDT in th Arctic sea and the Antarctic penguin, thousands of miles away from the areas where the pesticide is used. And because many of the disturbances which are introduced occur at vulnerable points such as the surface of the sea or on the sea floor, where oceanic life-supporting processes are generated, the dangers and magnified.

Although the modification of this microlayer has occurred only on the vicinity of oil spills and around the marine dump site of municipal wastes and no evidence has yet been found that detergents in the open ocean have significantly influenced the properties of the sea surface, the report points out that environmental catastrophies in certain coastal areas and seas might already have occurred.

Only improved management of the coastal areas and seas which involves controlling more closely the dumping of pollutants and careful monitoring of contaminant levels, will prevent a further deterioration of coastal environments. Much work however is still to be done before the several ecological sub-systems which constitute the ocean are fully understood. By the end of

1982 action plans were in operation or planned for 10 regions involving some 100 coastal States. Coastal and marine national parks and reserves were extended.

(e) Desertification

Desertification is now perceived as an intensifing world wide threat which, to varying degress, affects over 100 Member States of the United Nations. More than 20 percent of the world's population (850 million people) live in the 35 percent of the Earth's land surface (45 million km) which consists of the arid, semi-arid and sub-humid zones which is at risk of desertification. Three quarters of this area is already at least moderately desertified. Each year an additional 21 million hectares of agricultural land deteriorates through desertification to a point at which it is no longer economically productive. The last production has been valued at 26 billion dollars annually.

Although desertfication is exacerbate by severe drought, its principal cause is human over exploitation of dry lands through over-cultivation, overgrazing, poor irrigation practices and deforestation. When drought strikes the overtaxed livelihood systems, desertification results. The most devastating environmental and economic effects of desertification are in developing countries in the drylands. National balance of payments problems and local rural poverty lead to over-exploitation of lands and forests for food, fuel, cash crops and meat for export.

As these and other pressures increase to unsustainable levels, agricultural productivity and soil fertility decline, creating in turn new pressures which lead to further degradation. Recovery of land for agricultural purposes become costly and often economically unfeasible, thereby reducing the economic potential of the affected countries. In many of these countries agricultural land is also being reduced by salinization, urban sprawl and sand-dune encroachment. Increased desertification in one part of the world can indirectly induce desertification else where, as loss of agricultural land and increased food demand raise Prices and intensify pressures on marginal drylands in other food exporting and importing countries alike.

There are also other impacts on international trade, to the extent that developing countries affected by rapid desertification will have a reduced capacity to import goods and services from developed countries and from other developing countries. Other globally shared impacts include loss of valuable genetic resources, disruption of hydrological cycles and increase of atmospheric dust.

A great deal is already known about the causes and effects of desertification and about the measures needed to halt or combat it. Moreover, there is favourable cost benefit ration to justify, preventive action. The 1980 estimate for implementing the 1977 plan of Action to combat Desertification was 2.4 billion dollar annually over 20 years, which is only a fraction of the estimated value of agricultural production lost each year through desertification. It is, therefore, in the interests of all countries to establish some concreate cooperative actions between groups of developed and developing countries to halt further desertification, a goal which can be achieved, given a sufficient allocation of resources, on the basis of currently available knowledge.

(f) Loss of Tropical Forests

Although tropical forests are found in early half of all United Nations Member States, most are in twenty countries and nearly half their total area in three countries alone (Brazil, Indonesia and Zaire). The tropical forests are the world's richest biological zones, and are estimated to contain as much as 40 per cent of all terrestrial species on the planet. In addition, tropical forests provide a wide range of useful products of economic significance for both developing and developed countries. Undisturbed tropical forests are also home to millions of the world's tribal peoples. Many experts contend that tropical forests are being exploited at a rate and in a way that is ecologically destructive and economically unsustainable.

There are varying estimates of how much of the tropical forests will be lost by the year 2000 through complete conversion to other uses and by severe degradation. The results of the tropical forests resources assessment project conducted by FAO and UNEP indicated an estimated minimum loss of closed tropical forests of 12.5 per cent by the year 2000 (an average of 7.5 million hectares annually between 1980-2000) through complete conversion to other uses alone. Other studies which not only consider complete conversion but also include severe degradation caused, for example, by logging operations, by over-grazing, by over-exploitation for fuelwood, or by natural causes such as fires, insects and diseases, have estimated losses of tropical forests as upto 20 million hectares or more in a year.

The FAO/UNEP study estimates that tree planting on industrial and non-industrial plantations accounts for only 10 per cent of the area of tropical forests cleared annually.

The major causes of deforestation differ by region. According to the FAO/UNEP report in Latin America the two major causes are shifting cultivation and the conversion to pasture land for cattle ranching. In Asia and Africa, shifting cultivation by subsistence farmers and removals for fuelwood are major causes. The extraction of timber from tropical forests for industrial purposes is not considered by the FAO/UNEP reports as a major factor in deforestation. On a world-wide basis, the UNEP report estimates that fuelwood and charcoal removals from tropical forests are as much as eight times greater than for industrial wood. Serious degradation of loss of watershed forests in particular has a wide range of major ecological and economic effects through increased erosion, floods, landslides, and silting of hydro-electric facilities, irrigation systems reservoirs and harbours.

The lives and livelihood of half the world's population depend directly on the wise management of watershed forests and eco-systems. Deforestation also influences local, regional and possibly even global climatic conditions by altering hydrological regions, reducing vegetation cover and causing more heat and light to be reflected back to the atmosphere (the albedo effect), and through accelerated burning of the forest biomass, which contributes to carbon dioxide build-up in the atmosphere and the global warming trend. The economic, social and political implications of possible regional and global climate changes have only begun to be assessed.

(g) Soil Loss

Almost every environmental concern connects clearly with the importance of soil. Soil acts as a reservoir for water and other inorganic substances cycling through the eco-system. It provides habitat for many of the different populations or decomposer organisms that control the rate and capacity of nutrient recycling.

The importance of soil to agriculture can hardly be over emphasised. Agriculture provides about 44 percent of the poorest countries gross domestic product and 83 percent of their employment. Growing more food, building irrigation projects, conserving nature and precious wildlife and other needs must depend on proper soil management. The wide-ranging importance of soil confirms that soil management must consequently be many-sided. The broad range of natural and man-made stress that causes the two type of soil damage: physical displacement and alterations in soild structure and composition.

Uncontrolied shifting cultivation, salinization and water-logging flooding and plant cover destruction are same. An appropriate use of chemicals, weapons testing and deployment in warfare are some other casuses of soil nutrients and other types of soil degradation. Soil damage is also caused by patterns of dry land mountain side cultivation that carried out without arrangements for bunding and terracing to protect the topsoil.

Continuous cropping to meet growing food needs leave title time for soil conservation techniques. Deterioration of the productive capacity of the soil is most acute in areas affected by erosion, drought, mineral stress and other related factors. Soil resources that are vulnerable to these factors constitute 82 percent lands in South Asia.

While erosion is an integral part of the natural system, occurring even when land is covered by grass, or trees, its extent and intensity can be minimized provided land denudation is correspondingly reduced. The fact that it takes nature 100 years or more to build one inch of top soil should have engineered the vigilant observance of soil conservation measures. Instead, problems of soil erosion have continued ticking.

Countries in East and South Asia are particularly burdened with soil erosion occuring in watersheds. Anticipated acceleration of agricultural production and other activities between now and 2000 is casting its shadows: existing cultivable land will be further subjected to intensified use as cultivation of unsuitable areas persists; logging, firewood gathering and grazing will accelerate the elimination of vegetative cover, while the construction of roads, mine and water resources projects will be further pursued in the course of development. Such prospects place the least developed countries in an even more precarious situation. Slash and burn farming is a common enough culprit. Wielding destruction in gradual succession, it initially causes the loss of fertility by the thin topsoil, and later eroding its stability. In India, slash and burn farming is particularly rampant in the north-eastern and central regions affecting between six and eight million hectares of land.

Salinization affects 80 million hectares (8 percent) of total land in South Asia. Something should also be said about agricultural techniques which are designed to raise land productivity but instead lead to soil loss. The use of cheap nitrogen, for example, replaces nitrogens, fixing legumes, resulting in loss to topsoil. The commonality of problems, in addition to the implications transboundary soil management points up to the fact that soil problems cannot always be resolved within the context or national concerns. Clearly, what are required are mutual policies and strategies concerning land management, and the World Soils Policy provides an appropriate starting point.

Working in collaboration with the relevant UN bodies, UNEP has proposed the World soils Policy programme which endeavours to enhance international awareness, assist countries in the formulation of national soil policies, help develop technical and scientific knowledge and collect, compile and disseminate data on the use and management of the worlds soil resources.

(h) Depletion of genetic resources

25,000 species and more than 1,000 vertebrate species and sub-species are now estimated to be threatened with extinction, yet these two categories of organisms represent only a small proportion of the terrestrial species on the planet. According to some estimates as much as from half a million to a million species would be extinguished over the next two decades. Extinction of species on this compressed scale would be without precedent in human history. As much as two-thirds of all terrestrial species and the majority of endoangered species are located in developing countries.

A significant loss of species in developing countries would have a world wide effect on future agriculture, industry and science, and on human health and welfare generally. The most serious threat to many species is the disruption and destruction of their habitat by man, especially in the tropical forests which alone harbour as much as 40 per cent of all terrestrial species. Significant species loss could directly affect human health and welfare locally, and even nationally, in many developing countries, where wild plants and animals are a source of fodder, traditional medicines and income especially for the poor in rural areas. In some countries tourism largely on wildlife is a major source of foreign exchange.

In many developed countries over one-thrid of the gain in the productivity of many major agricultural crops over the last few decades has been attributed to breeding programmes using genetic material from related wild plant varieties, many of which are found in developing countries. Moreover, many of the major crops in developed countries now have a limited genetic base. For example, more than half the Canadian prairie wheatlands are devoted to a single variety, Neepawa, which was developed with germplasm from Kenya; 72 per cent of United states potato production depends on only four varieties; and the entire United States soyabean industry is derived from a mere six plants from Asia. A significant loss of relavant wild species could limit the possibilities for maintaining or increasing the yields of major crops, such a wheat, maize, rice potatoes and for combating new strains of diseases or pests which threaten them.

Genetic stocks from one part of the world frequently make an enormous contribution to another. Current research shows that enormous opportunities lie ahead, such as the utilization of Asian blue-green algae of fix nitrogen for crop fertilization, which may constitute a gain of hundreds of millions of dolars a year to the economics of developed countries alone. Major medicines and other pharmaceutical products in developed countries are also often dependent on plant and animal species. For the United States of America it has been estimated that more than 40 per cent of the prescriptions written each year contain a drug of natural origin-from higher plants (25 per cent), microbes (13 per cent) or animals (3 per cent) as the sole active ingredient or as one of the main ones. It has been estimated that among the 10 percent of species that might be extinguished by the year 2000 there could be many plants with potential for use in new plantbased drugs and pharmaceuticals with a combined world wide commercial sales value of as much as 200 million pounds annually.

Biomedical research in developed countries is also vitally dependent on wild species, many of which are threatend with extinction. The innovative capacity and future of the rapidly expanding field of bio-technology, practised largely in developed countries, relies considerably on genetic patterns and resources from many wild plants and animals as only about 10 per cent of the world's plant species and 1 per cent of the animal, species have so far been subjected to even a preliminary screening for potential medical or commercial applications, a significant loss of species could also represent a permanent loss of sources for more effective chemicals, medicines and other useful products.

In spite of this interdependent relationship, species loss continues to accelerate. Developing countries often lack the capacity and economic incentive for extensive species protection. They have derived little or no economic benefit from the development and use of genetic materials in developed countries, where the scientific and technical expertise to exploit species and their genetic resources predominantly resides. Nevertheless, developing countries are and will remain the principal source of genetic materials from wild plants and animals. This clearly shows that there is a common incentive for developed and developing nations to prevent a potentially catastrophic loss of species. The mutual advantage of comprehensive joint effort could result in new medicines and other useful products that would contribute to improved levels of human health and welfare in all countries.

Human progress has always depended on our technical ingenuity and a capacity for cooperative action. These qualities have often been used constructively to achieve development and environmental progress: in air and water pollution control, for example, and in increasing the efficiency of material and energy use. Many countries have increased food production and reduced population growth rates. Some technological advances, particularly in medicine, have been widely shared.

But this is not enough. Failures to manage the environment and to sustain development threaten to overwhelm all countries. Environment and development are not separate challenges: they are inexorably linked. Development cannot subsist upon a deteriorating environmental resource base: the environment cannot be protected when growth leaves out of account the costs of environmental destruction. These problems cannot be treated separately by fragmented institutions and policies. They are linked in a complex system of cause and effect. First, environmental stresses are linked one to another. For example, deforestation by increasing run off, accelerates soil erosion and siltation of rivers and lakes. Air pollution and acidification play their part in killing forests and lakes. Such links mean that several different problems must be tackled simultaneously. And success in one area, such as forest protection, can improve chances of success in another ares, such as soil conservation.

Second, environmental stresses and patterns of economic development are linked one to another. Thus agricultural policies may lie at the root of land, water, and forest degradation. Energy policies are associated with the global greenhouse effect, with acidification, and with deforestation for fuelwood in many developing nations. At these stresses all threaten economic development. Thus economics and ecology must be completely integrated in decision-making and law making processes not just to protect the environment, but also to protect and promote development. Economy is not just about the production of wealth, and ecology is not just about the protection of nature: they are both equally relevant for improving the lot of humankind.

Third, environmental and economic problems are linked to many social and political factors. For example, the rapid population growth that has so profound an impact on the environment and on development in may regions is driven partly by such factors as the status of women in society and other cultural values. Also, environmental stress and uneven development can increase social tensions. It could be argued that the distribution of power and influence within society lies at the heart of most environment and development challenges. Hence new approaches must involve programmes of social development, particularly to improve the position of women in society, to protect-vulnerable groups and to promote local participation in decision making. Finally, the systemic features operate not merely within but also between nations.

National boundaries have become so porous that traditional distinctions between matters of local, national, and international significance have become blurred. Ecosystems do not respect national boundaries. Water pollution moves through shared rivers, lakes, and seas. The atmosphere carries air pollution over vast distances. Major accidents—particularly those at nuclear reactors or at plants or warehouses containing toxic materials—can have widespread regional effects. Many environment-economy links also operate globally.

For instance, the highly subsidized, incentive driven agriculture of industrialized market economies generates surpluses that depress prices and erode the viability of the often neglected agriculture of developing countries. Soils and other environmental resources suffer in both systems. Each country may devise national agricultural policies to secure short-term economic and political gains, but no nation alone can devise policies to deal effectively with the financial economic, and ecological costs of the agricultural and trade policies of other nations.

INDEX

Population laboratories, 216